30-SECOND ENGINEERING

30-SECOND ENGINEERING

50 key fields, methods and principles, each explained in half a minute

Editor
James Trevelyan

Contributors
Roma Agrawal
John Blake
Colin Brown
George Catalano
Doug Cooper
Kate Disney
Roger Hadgraft
Jan Hayes
Marlene Kanga
Gong Ke
John Krupczak
Raj Kurup
Julia Lamborn
Andrew McVeigh
Seán Moran
Paul Newman
Hung Nguyen
Jenn Stroud Rossmann
Veena Sahajwalla
Tomás A. Sancho
Jonathan Scott
Tim Sercombe
Paul Shearing
Donglu Shi
Matthew L. Smith
Jorge Spitalnik
Neill Stansbury

Illustrations
Nicky Ackland-Snow

First published in the UK in 2019 by
Ivy Press
An imprint of The Quarto Group
The Old Brewery, 6 Blundell Street
London N7 9BH, United Kingdom
T (0)20 7700 6700 **F** (0)20 7700 8066
www.QuartoKnows.com

British Library Cataloguing-in-Publication Data
A catalogue record for this book is available from the British Library.

ISBN:978 1 78240 837 6

This book was conceived, designed and produced by
Ivy Press
58 West Street, Brighton BN1 2RA, UK

Publisher **Susan Kelly**
Creative Director **Michael Whitehead**
Editorial Director **Tom Kitch**
Art Director **James Lawrence**
Project Editor **Katie Crous**
Designer **Ginny Zeal**
Illustrator **Nicky Ackland-Snow**

Cover images: Ovidiu Leanca (M); DidGason (L);
Ivy Press/ Adam Hook (B)

Printed in China

10 9 8 7 6 5 4 3 2 1

CONTENTS

INTRODUCTION

James Trevelyan

Safety is the primary concern for engineers. When accidents occur, engineers learn from mistakes and improve standards as a way of passing on experience.

Almost everything we see, feel, touch, eat or drink entails engineering: engineering is fundamental for human civilization, and engineers are the people with the core technical ideas and knowledge.

Yet, engineering itself is a mysterious profession. Many imagine engineers designing and performing complicated mathematical calculations. Some engineers do that, but very few spend much time on it. Others think that engineers build bridges and make cars. However, few engineers would know how to fix a car, let alone work with tools on a bridge. Of course, locomotive engineers drive trains in America, but in this book you will read about engineers as members of a knowledge-based profession.

The best way to appreciate engineering is to understand what engineers really do, and recent research has greatly expanded our knowledge of this. Even with the enormous variety that comes with the near-300 specialized engineering fields that exist today, there is remarkable similarity in the work of engineers, everywhere. Most engineers rely on a handful of common ideas and methods, explained in the first section of this book.

So, what do engineers do? In a few words, they are people with technical knowledge and foresight who conceive, deliver, operate and sustain man-made objects and systems that enable people to do more with less effort, time, materials, energy, uncertainty, health risk and environmental disturbances.

Organizing their work into specific projects, in the first phase, engineers conceive safe solutions for human needs and predict how well these solutions will operate and the cost to build, operate, sustain and, eventually, remove them. There are always uncertainties: engineers inform investors about risks and consequences. Sufficient trust and confidence need to be gained before investors are willing to commit

finance for project execution, the second phase, long before benefits from the project will arise.

Often working in large teams, engineers plan, organize and teach people to purchase and deliver components, tools and materials, and then transform, fabricate and assemble them to deliver the intended solution. They work with an agreed schedule and budget, handling unpredictable events that influence progress, performance, safety or the environment. Later, they organize sustainment: operations, upgrades, maintenance and repairs. In the final phase, engineers plan and organize removal and disposal, environmental restoration, reuse, refurbishment or even recycling of the materials.

Expectations need to be satisfied well enough for investors to come back and commission more projects. Most of an engineer's time is spent coordinating collaborative efforts by skilled people, guided by shared technical knowledge. Engineering successes often reflect the combined performances of tens (or even hundreds) of engineers and thousands of other people worldwide, building on decades of experience. Technical work, such as predicting performance, designing and solving problems, takes up much less time.

Certainty is impossible with unpredictable activities by so many people combined with natural variations in materials and the environment. Yet engineers have developed systematic methods that provide amazing predictability. A century ago, few people watching the hair-raising exploits of aviation pioneers could have imagined the amazing safety and reliability of modern air travel.

Invention and innovation is a fundamental ethos in engineering, tempered by accumulated knowledge and standard methods shaped by past experiences. Another guiding principle is the ethical duty not to

Engineers strive to create highly reliable products, and endure great uncertainty and anxiety when testing them in extreme conditions.

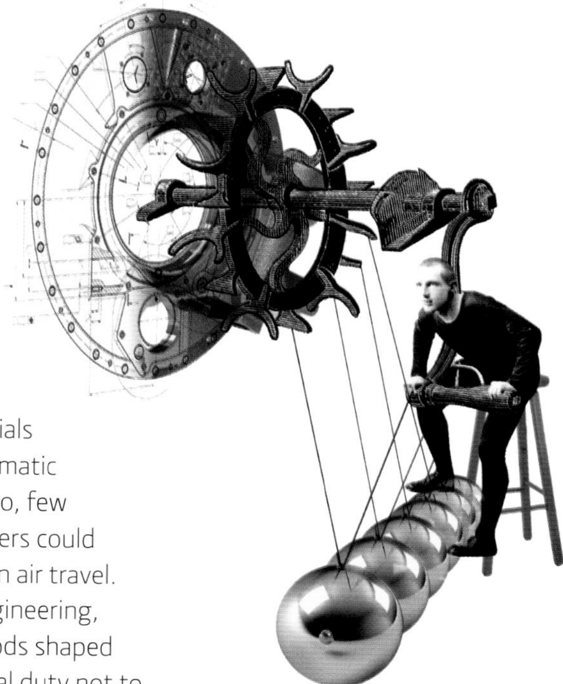

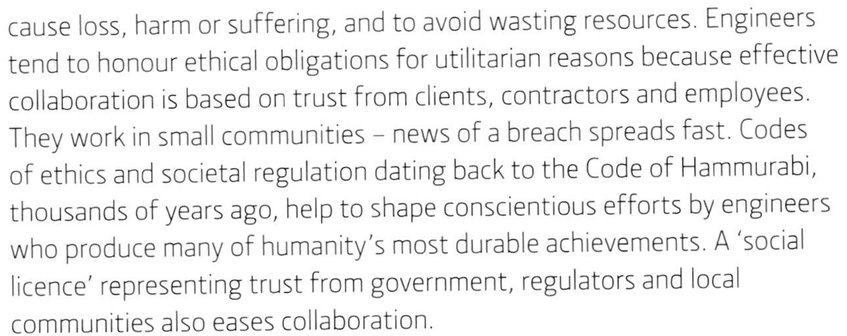

cause loss, harm or suffering, and to avoid wasting resources. Engineers tend to honour ethical obligations for utilitarian reasons because effective collaboration is based on trust from clients, contractors and employees. They work in small communities – news of a breach spreads fast. Codes of ethics and societal regulation dating back to the Code of Hammurabi, thousands of years ago, help to shape conscientious efforts by engineers who produce many of humanity's most durable achievements. A 'social licence' representing trust from government, regulators and local communities also eases collaboration.

Traditionally a culturally diverse male-dominated profession, women are gradually making their presence felt, especially in fields like biomedical, environmental, food processing and chemical engineering. Many firms recognize the value of diversity and are now actively seek to recruit and retain female engineers, and are creating more inclusive workplaces.

Improved sensing technology helps create public awareness of pollution, creating social pressure on companies that enables engineers to create better solutions.

Some have argued that future computers with artificial intelligence (AI) will perform many of today's engineering jobs. So far, the main results from AI have been to improve IT system performance, providing engineers with faster access to appropriate information and enabling more effective robots. Forecasts that robots and AI would eliminate factory work, for example, proved to be premature. Yet computer systems, often with AI components, do help to enormously extend human capabilities.

Engineers will continue to provide essential leadership for many of the greatest advances in human civilization. Most engineers have high levels of job satisfaction and often enjoy the thrill that comes from spending large amounts of money to transform ideas into reality, bringing great benefits to vast numbers of people.

Engineers are creating more energy-efficient high-speed trains that enable more people to travel further and faster using renewable energy sources.

How this book works

30-Second Engineering is your essential guide to engineering, covering the pivotal developments and feats that have created the world in which we live today. From establishing civilization and all its necessary amenities, through taming the forces of nature to our advantage, to building a future of innovation and sustainability – engineering has always been at the forefront of change. Experts from around the globe share their knowledge and guide us through the carefully selected topics, to ensure all fundamental bases are covered, clearly and concisely. Each aspect of engineering is presented on a single page, alongside an illustration that reflects key themes. The main paragraph, the **30-Second Foundation**, expands on the **3-Second Core**: the topic in one or two sentences. The **3-Minute Idea** takes it further, adding intriguing aspects of the subject. Each chapter also features a biographical profile of a widely respected engineer – the men and women whose lives tell us something about the lives of all engineers. The book begins with an overview that explains what engineers do, their contributions and techniques common to all disciplines. It then delves into engineering disciplines, their associated achievements and goals: we learn where civil and environmental engineering started and where it is taking us; mechanical, materials and mechatronic engineering introduces us to motion and energy; we compare energy sources and how they are harnessed; and through electrical engineering we glimpse into the future of automation, technology and transport. The final chapter addresses the challenges engineers face today to build a better world for all of us tomorrow.

ENGINEERING METHODS

ENGINEERING METHODS
GLOSSARY

boundary element method Calculation of stress, pressure, temperature, displacement, electric or magnetic fields throughout a component, using conditions at the component boundary. This can be much faster than the finite element method for certain problems, such as simulating adhesive contact between parts.

convergence Many engineering calculations start with an estimated solution and then refine the solution, step by step, converging on the solution with increasing accuracy. Sometimes the accuracy does not improve, however, and a different method has to be found.

eddy current Movement of a solid conductor relative to a magnetic field causes recirculating currents in the conductor, generating a magnetic force that opposes the motion.

finite element method Calculation of continuously variable properties, such as stress, pressure, temperature, displacement, electric or magnetic field strength, throughout a component, by subdividing the component shape into a large number of small elements, each of which can be analysed with well-known equations. The method originated with collaborations between aircraft designers and mathematicians who had access to early digital computers in the 1950s.

force External influence that causes a stationary object to move, or a moving object to slow down, change direction or accelerate. An example is gravity: the force attracting bodies towards the centre of the earth. Another example of force is the thrust on an aeroplane resulting from its engines.

heuristic, or rule of thumb Approximate mathematical or qualitative relationship that engineers use to make predictions when detailed measurements or mathematical analysis cannot be used, often because insufficient time, understanding or data is available for more accurate methods to be used.

model Set of mathematical equations, usually embodied in a computer program or spreadsheet, which engineers use to predict the behaviour of an engineering system. Engineers also use physical models, often scale models with identical proportions but a different size relative to the system being studied.

network Many systems can be analysed as networks consisting of separate elements (or nodes) joined together by a finite number of connections. Examples include electronic circuits, telephone networks, and networks of pipes, pumps and tanks.

stability Tendency of an engineering system to remain in a relatively unchanging state. As a system approaches the limit of stability, smaller disturbances can cause a sudden and usually undesirable, often uncontrollable change in conditions. For example, a slope can remain stable, even in an earthquake, unless heavy rain has reduced its stability by reducing friction forces between tree roots, rocks and soil. Under those conditions, even a small tremor can cause a landslide.

stakeholder Individual or group of people who can influence, or could be influenced by, an engineering activity.

strain Relative deformation caused by stress. Tension stress causes elongation, and compression stress causes squeezing of material. Shear stress causes layers of material to slide relative to each other.

stress Applied force over a given area inside a body. Usually, an external force results in an uneven distribution of stress inside a body. Stretching or fracture starts where stress exceeds the material strength. Stress can be tension (pulling apart), compression (squeezing together) or shear (opposed forces on opposite sides of material).

system and systems thinking Engineers think in terms of systems: complex collections of components that interact with each other and their surroundings. Engineers construct an artificial boundary enclosing all the components in a system, and classify interactions that cross the boundary as 'system inputs' and 'system outputs'. Inputs are external changes that cause changes inside the system; outputs are changes inside the system that cause changes outside. Engineers often subdivide systems into a hierarchy of sub-systems until the individual components are simple enough to analyse separately.

DIVIDE & CONQUER

30-second foundation

Engineers predict the behaviour of systems so complex they can defy human comprehension. They divide complex systems into simpler parts with carefully chosen boundaries and account for the influences that cross boundaries. For instance, when predicting a car's behaviour, engineers draw a 'free body diagram' for each wheel. The diagram excludes everything but the wheel and the forces that act on it. It shows the road force pushing the wheel up. The car's weight, transmitted to the wheel through the axle, acts downwards. The axle driving torque twists the wheel and the road exerts a friction force. The wheel and tyre could be defined separately, and subdivided into many smaller elements, with analysis of each one. Finer subdivision can yield more accuracy, but it takes engineers time to define the element boundaries and forces. Computers handle calculations, but the engineer has to guide element definition and assess the accuracy and reliability of calculated results. Engineers learn appropriate levels of subdivision depending on the accuracy required and the consequences of simplification errors. They use this approach for modelling complex systems, such as the structural stability of an underground mine, rush-hour traffic congestion and radar beam forming.

3-SECOND CORE

Engineers analyse complex systems by dividing them into simpler elements and accounting for influences that cross element boundaries. Computers do most of the calculations.

3-MINUTE IDEA

A complex piping network carrying water is analysed in parts. At each pipe joint, the sum of the separate water flows towards the joint must be zero: water cannot disappear without a leak. The pressure change along the pipe depends on the pipe elevation and flow. We can represent all these relationships with simultaneous equations that a computer can solve. Electrical circuits, factory production lines, city traffic congestion and communication networks can be analysed using similar methods.

RELATED TOPICS

See also
APPLYING MATHEMATICS
page 16

ENGINEERING THINKING
page 26

FORCE EQUILIBRIUM PRINCIPLE
page 38

3-SECOND BIOGRAPHIES

CHARLES-AUGUSTIN DE COULOMB
1736–1806
French engineer who described friction, explaining why cars skid during braking, and the electrostatic forces so critical in nanotechnology.

GABRIEL VOISIN
1880–1973
French aviation pioneer who invented the first anti-lock brakes to help prevent skidding.

30-SECOND TEXT

James Trevelyan

Automotive engineers design new cars using components that have proved to be reliable.

A
B
A

APPLYING MATHEMATICS

30-second foundation

3-SECOND CORE

Mathematics is most often applied by engineers indirectly through the use of large software packages. Predicting damage from an automobile collision is one of many engineering applications of maths.

3-MINUTE IDEA

Computational fluid mechanics (CFD) software solves complex differential equations that describe fluid flow. Beautiful images often appear when the results are portrayed graphically. CFD is not necessarily accurate in all situations. Numerical methods sometimes fail to converge to a usable solution. CFD software has been refined for particular applications where experimental data can confirm accuracy – in the aircraft industry, for example.

Frequently applying their maths indirectly, engineers rely on software with built-in analysis and also a tacit understanding of maths principles. They need instinctive ability to choose appropriate computational methods for given circumstances. Early in an engineering project, fast and rough calculations may be best, because an answer within 20 per cent is often sufficient. At this stage, engineers and project owners are usually deciding whether the project is worth investigating in detail and which alternative engineering approaches should be investigated. Later, greater precision is needed for critical aspects of the project: an engine designer may be searching for the most efficient layout of passages for the air and fuel mixture to enter the cylinders of an engine; others may use analysis software to predict how well a new car design will protect passengers in a collision with another vehicle. These computer analyses take time, so engineers need to understand cost and limitations. Usually, the software is operated by specialist engineers who know how to set up the appropriate mathematical models quickly. They need to check that the results are reasonable, often using simpler approximate calculations that can be set up in an Excel spreadsheet or even calculated manually.

RELATED TOPICS

See also
DIVIDE & CONQUER
page 14

BEYOND SCIENCE
page 18

SIGNAL PROCESSING
page 108

3-SECOND BIOGRAPHIES

SIR ISAAC NEWTON
1642–1726
Laid the mathematical foundations that underpin modern engineering.

JOSEPH FOURIER
1768–1830
Pioneered methods for predicting heat conduction through solid bodies.

KATHERINE JOHNSON
1918–
Prepared orbit predictions for the first manned US space flights.

30-SECOND TEXT

James Trevelyan

Mathematics helps engineers predict aircraft performance, reducing the need for prototype flight tests.

BEYOND SCIENCE

30-second foundation

Engineers use experience and scientific knowledge to predict the behaviour of objects that only exist as sketches. Sometimes they have measurements from scale models to compare with predictions and they investigate anomalies. They often find themselves far ahead of science. For example, gears have been used for thousands of years, yet comprehensive mathematical theories have only recently emerged. Boiling and condensation of refrigerant gas in an air conditioner is almost impossible to predict accurately from current scientific knowledge. Therefore, engineers use estimates from experience and simple calculations. They test a prototype in climate chambers that simulate the actual conditions in which it will be used. They add or remove gas slowly until they judge that the cooling performance provides the best compromise between power consumption and cooling effect in common climatic conditions. Here we see the essential distinction between science and engineering: engineers work with science to the extent possible with today's knowledge (and software). They predict the behaviour of systems that don't yet exist, so confirmatory measurements have to come from models. With time and resource constraints, they often rely on experience and practical heuristics.

3-SECOND CORE

Engineers often have to work beyond the limits of scientific knowledge. They conduct experiments with scale models and use experience as a guide.

3-MINUTE IDEA

Physics and mathematics provide most of the scientific knowledge used by engineers – mechanics, thermodynamics, electromagnetics, atomic and molecular structure of materials, and material properties. Well-used mathematical methods include calculus, matrix algebra, infinite series and probability. Chemistry contributes to material properties knowledge and helps to explain material degradation. Engineers draw on the life sciences, for food processing, packaging and storage, removing pollution, making artificial body parts and manufacturing medicines.

RELATED TOPICS

See also
PROBLEM-SOLVING DESIGN
page 20

ENGINEERING THINKING
page 26

NANOTECHNOLOGY
page 104

3-SECOND BIOGRAPHIES

BLAISE PASCAL
1623–62
Clarified concepts of pressure and helped invent the mechanical calculator.

ROBERT HOOKE
1635–1703
Described the theory of elasticity and coined the biological term 'cell'.

LÉON FOUCAULT
1819–68
Invented a pendulum to prove the Earth's rotation and discovered eddy currents.

30-SECOND TEXT

James Trevelyan

Science is a foundation for engineering, yet engineers often devise ideas without scientific understanding.

PROBLEM-SOLVING DESIGN

30-second foundation

Engineering design is a multi-faceted, problem-solving effort leading to creation of products or systems that meet the needs of end users or customers. The first step is identification of the requirements and conditions that the system must satisfy, which requires consideration of the needs of all stakeholders. Besides the end user, stakeholders can include manufacturers, distributors, service and repair technicians, sales people, purchasing agents and government regulators. Embodying the desired functions into physical form through components that provide specific functions and solve sub-problems creates a technological system. Carrying out the engineering design process requires having knowledge of available components, component functions and an ability to envision the functions that could be provided by a particular form. Quantitative analysis is often used to determine specific component parameters and to match the inputs and outputs of components internal to the system. In designing a technological system, more than one solution is possible since there are many components or ways to embody particular functions. The design is compared to the requirements for performance, characteristics and features. This may involve testing of tentative or prototype designs.

3-SECOND CORE

Design is a problem-solving process to determine the needs of the user and develop efficient creative solutions. Designers combine individual components into overall systems through analysis, testing and iteration.

3-MINUTE IDEA

Today, 'design thinking' is a term being used to describe the creative approaches used in the process of designing, as well as the methods used to match the needs of customers, with what is achievable at any current state of technology and business. Hallmarks of design thinking include striving for empathy with the user and the use of multiple perspectives as a way to discern user needs, resolving competing or contradictory factors and envisioning innovative approaches.

RELATED TOPICS

See also

STANDARDS & SPECIFICATIONS
page 22

ENGINEERING THINKING
page 26

ENGINEERS & ARCHITECTS
page 42

3-SECOND BIOGRAPHIES

HERBERT A. SIMON
1916–2001

Nobel-prize winning economist whose 1969 book *The Sciences of the Artificial* was one of the first to analyse the design and problem-solving process.

DAVID M. KELLEY
1951–

A member of the US National Academy of Engineering, Kelley cofounded world-leading design firm IDEO in 1991.

30-SECOND TEXT

John Krupczak

Frequently, redesign or iteration is required to achieve a satisfactory result.

STANDARDS & SPECIFICATIONS

30-second foundation

Scientific knowledge and experience guide the work of engineers. Personal experience is only a small part; standards incorporate experience acquired by generations of engineers. They provide fast and convenient design and calculation methods that provide safe and reliable results. Standards are mainly issued by national, professional and industry-based organizations. A group of railway engineers formed the American Society for Testing Materials (ASTM) in 1902 to standardize steel testing; now ASTM provides standards and training worldwide. The International Organization for Standardization (ISO) co-ordinates global standards. Standards evolve as technology changes and engineers learn more, sometimes from mistakes. Engineers write specifications and create drawings to define products so that components provided by different companies will fit together and perform as expected. Specifications come in two broad types: test and method. The former defines tests or inspections to decide if a product is acceptable. It can be hard to tell from a test if the product will still perform after 30 years, so a specification may also define fabrication methods. To save time, specifications refer to standards for much of the technical detail.

3-SECOND CORE

Standards provide convenient and safe working methods based on experience acquired by generations of engineers. Specifications and drawings define products so that the components will work as expected when assembled.

3-MINUTE IDEA

Specifications are interpreted differently, depending on local expectations. A product accepted in one industry may be rejected in another, just because a wire has a different colour. A machine might be accepted even though it does not meet all performance requirements, because specifications are considered 'aspirational' goals. Eventually, similar machines off the same production line achieve most of the stated performance, and earlier models can be upgraded.

RELATED TOPICS

See also
PROBLEM-SOLVING DESIGN
page 20

ENGINEERING THINKING
page 26

DEFENCE READINESS
page 64

3-SECOND BIOGRAPHIES

GASPARD MONGE
1746–1818
Invented descriptive geometry, the basis of technical drawings.

PIERRE ÉTIENNE BÉZIER
1910–99
Created techniques and software to represent curves and surfaces.

DOUGLAS TAYLOR ROSS
1929–2007
Developed software to help engineers perform design calculations.

30-SECOND TEXT

James Trevelyan

Standards help ensure that earlier engineers' experience guides future generations.

Y054
Y100
Y050

MAKING IT HAPPEN

30-second foundation

About 30 per cent of an engineer's time is spent negotiating the willing and conscientious collaboration of other people within an agreed time schedule. They rely on others to contribute knowledge, experience and skills; the expertise needed for engineering is distributed amongst the people who collaborate to make it all happen. A network of social relationships creates the trust needed for this collaboration: social interactions lie at the core of technical practice. Engineers learn much from others – colleagues, suppliers, contractors, skilled artisans, technicians, financiers, lawyers, end users and the local community. This knowledge is tacit or implicit – seldom written down. Relying on human memory and complex social interactions, there is unpredictability beyond variability in materials, natural situations and weather. Most engineering failures are communication failures. An engineer has to make sure that everyone has sufficient understanding of essential features for appropriate implementation by others. While this social complexity is challenging, it is more so in developing societies, where trust can be elusive. Combined with the tenuous presence of knowledgeable engineering suppliers, this explains why it is more costly to achieve comparable results in engineering enterprises.

3-SECOND CORE
Engineers spend much of their time organizing collaboration by all the people needed to construct and deliver the products of engineering work, providing their specialized skills and knowledge.

3-MINUTE IDEA
While many people think that engineers build things like bridges and cars, very few actually make or build anything as part of their professional work. Instead they organize construction and delivery by other people, such as technicians, who are far more skilled with their hands and mechanized tools. Highly skilled technologists using computers create the drawings often associated with engineers. Those who enjoy hands-on work mostly do it in their spare time as hobbies.

RELATED TOPICS
See also
ENGINEERING THINKING
page 26

ORGANIZATIONAL SAFETY
page 82

3-SECOND BIOGRAPHIES

FREDERICK WINSLOW TAYLOR
1856–1915
American mechanical engineer known for the influential early theory of scientific management, also made many improvements in steel making.

LILLIAN EVELYN GILBRETH
1878–1972
American psychologist and engineer, first female member of the American Society of Mechanical Engineers, who developed time and motion study methods that led to enormous productivity improvements in factories.

30-SECOND TEXT
James Trevelyan

Engineering relies on an invisible web of trusting interpersonal relationships to make things happen.

ENGINEERING THINKING

30-second foundation

Engineers use judgement and experience to find solutions for human needs, examining existing solutions to identify feasible improvements. They think about systems interacting with people and the environment. Parts of systems interact with each other; each part can be understood as a system itself. Visualizing helps an engineer transition from the abstract to reality, mentally rehearsing the use of materials and space to create practical solutions. Abstract thinking embraces invisible effects such as stress, electric and magnetic fields or thermal conductivity, predicting how a concept might work. Sketches and 3D prototypes follow, and engineers iterate and optimize a solution within constraints such as lifecycle cost, available time and stakeholder needs. Discussions with peers, suppliers, customers and technicians help engineers shape their ideas. Successful solutions tend to emerge from sharing collaborations that learn from failures, responding to feedback. Engineers measure system performance to confirm expectations, deliberately inducing failure by going outside the 'design envelope'. Measurements help shape engineers' intuitive ideas on how systems and their worlds actually work. Unexpected results often force a rethink, strengthening an engineer's mental model.

3-SECOND CORE

Engineering thinking requires abstract visualization with as much discussion and debate as practical implementation. Data from measurements serve as the ultimate reality check.

3-MINUTE IDEA

Good engineers appreciate their own limitations and build collaborative relationships with people who complement their skills. They enjoy the satisfaction of seeing their ideas used, creating tangible benefits for people. They also gain satisfaction from resolving apparently conflicting requirements with solutions admired by peers. They value opportunities to work on humanitarian ventures.

RELATED TOPICS

See also
BEYOND SCIENCE
page 18

PROBLEM-SOLVING DESIGN
page 20

ENGINEERS & ARCHITECTS
page 42

3-SECOND BIOGRAPHIES

ARCHIMEDES
287–212 BCE
Measured the volume of irregular objects to distinguish real gold from alloys.

LEONARDO DA VINCI
1452–1519
Invented a computer, solar collector, helicopter and armoured vehicle.

MATTHEW BOULTON
1728–1809
Saw the critical importance of precision manufacture for making efficient steam engines.

30-SECOND TEXT
Colin Brown

Measurements and observed real world performance often force engineers to rethink their ideas.

11 January 1843
Born in Gravelmount, County Meath, Ireland

1861
Completes apprenticeship at Bagnell & Smith; becomes Assistant Engineer

1865
Arrives in New Zealand, surveys route for Arthur Pass road

1868
Appointed District Engineer, Greymouth; completes harbour wall, railway and other projects

1872
Becomes Canterbury District Engineer, Christchurch

1875
Appointed West Coast District Engineer, Hokitika

1880
Becomes Inspecting Engineer, Dunedin; admitted to Institution of Civil Engineers, London

1883
Appointed Under Secretary Public Works, Wellington

1891
Appointed General Manager Railways and Engineer-in-Chief for Western Australia

1896
Completes design and estimates for Goldfields water supply pipeline

1898
Funding approved for pipeline

1900
Fremantle Harbour completed

10 March 1902
O'Connor dies by suicide, pipeline completed on (amended) schedule

CHARLES YELVERTON O'CONNOR

Born in Ireland in 1843, O'Connor set his heart on becoming a civil engineer from a young age. He learned railway construction apprenticed with engineering firm Smith and Bagnell. The great famine devastated Ireland, so O'Connor migrated to New Zealand in search of work. After surveying roads through the mountainous South Island terrain, enduring continual downpours and flooding, he was soon constructing infrastructure for mining projects. He gained respect for firm handling of corrupt contractors, building harbours at Greymouth, Westport and Hokitika, where shifting sands and shingle provided challenging conditions. The remote location honed his abilities to organize supplies of machinery and materials from Britain and Europe, months away by sea.

O'Connor was sought out by John Forrest, Premier of Western Australia, to build ports and water supplies for mining rich gold deposits in desert locations. His New Zealand experience was critical: Fremantle Harbour was constructed on shifting sand bars near the mouth of the Swan River. O'Connor incurred the wrath of contractors unaccustomed to his insistence that they meet all their undertakings before being paid. They retaliated by influencing powerful relatives to harass O'Connor with derisory newspaper articles.

O'Connor's greatest project was a pipeline to convey water from a dam near Perth to Kalgoorlie, 580 km (360 miles) inland, in a parched salt desert. O'Connor proposed an 80-cm (30-in) diameter pipeline to be constructed in one of the world's remotest locations with little industrial capacity. Anticipating apprehensions of risk averse London financiers, he designed his pipeline in 14 sections. Far longer and larger than any yet built, he presented it as 'simply a repetition, several times over, of schemes within the knowledge of most engineers'.

Financing took time, and a natural rock fault under the main storage dam delayed work by 12 more months. In 1901, Premier Forrest was elected to parliament in Melbourne, leaving O'Connor without strong local political support. Corruption accusations, mounting criticism from newspapers and impatient politicians, and stress from managing several 'mega-projects' simultaneously led to O'Connor committing suicide in 1902. His pipeline was completed eight months later only 10 per cent over the original budget. The pipeline, harbour and railways built by O'Connor transformed Western Australia, creating immense prosperity, and are still in use today.

James Trevelyan

CIVIL & ENVIRONMENTAL ENGINEERING

CIVIL & ENVIRONMENTAL ENGINEERING
GLOSSARY

axes Defined directions in space, often with *x* and *y* axes defining horizontal reference directions at right angles (perpendicular) to each other, and a *z* axis defining the local vertical direction.

conservation laws Fundamental laws of physics that guide engineers; include conservation of mass and energy. Mass cannot be created or destroyed, and energy cannot be created or destroyed. Nuclear physics tells us otherwise, but these laws are sufficiently accurate for almost all engineering purposes.

coordinate system A set of three intersecting mutually perpendicular axes. The point at which all three axes intersect is called the 'origin'. Engineers locate points in three-dimensional space with coordinates that define position along each axis relative to the origin. Engineers may use many different coordinate systems for different parts of a structure or system.

equilibrium A state in which forces (or other influences) on a system or component are balanced, so there is no tendency to move.

formwork Temporary mould for creating concrete structures.

foundation (structural) Specially designed parts of a building structure that transfer loads to the underlying soil or rock.

geomechanics Physical principles governing behaviour of soil and rock; core knowledge for geotechnical engineering.

geotextile Fabric material for reinforcing loose soil or gravel, preventing erosion.

load (civil, mechanical engineering) Can refer to a force or stress. 'Dead' loads are constant forces on a structure due mainly to gravity. 'Live' loads are variable forces caused by influences such as people, vehicles, wind, earthquakes, etc.

load path The pattern of stress in a structure. Load paths can change due to settlement, for example.

o-ring (mechanical engineering) Rubber sealing ring located in specially machined grooves.

pile (civil engineering) Large steel, reinforced concrete, occasionally wood rod or tube, placed, pushed or hammered into soil to provide additional strength for foundations, or to transfer loads directly to rock underlying soft soil.

pollution Harmful materials in water, soil, air or vegetation.

plant (civil, mechanical engineering) Machinery used in manufacturing or other engineering activity. Mobile plant refers to special vehicles such as excavators.

reinforced concrete Concrete containing small stones poured into formwork around an array of twisted or ribbed steel bars and wire, forming a strong composite material used for buildings and bridges. Most steel bars are placed selectively at the locations where the concrete must withstand tension loads.

service Pipe or cable connection enabling the supply of fluid or energy to an engineering process or for general use; also refers to maintenance activity.

site remediation Removal or stabilizing of pollution at a site of engineering or other human activity to prevent harmful effects of pollution.

slurry Crushed rock, powder, silt or sand mixed with a liquid (usually water) so it can be transported by pumping through a pipeline.

settlement (geotechnical engineering) Gradual sinking of foundations caused by the weight of the structure acting on underlying soil.

survey Systematic measurements collected to provide data for planning construction or other engineering work.

tailings Waste material from mining operations.

turbine Wheel with blades designed to transform kinetic energy in a flowing fluid into mechanical energy in a rotating shaft, often to turn a generator to transform mechanical energy in to electricity.

CIVIL ENGINEERING

30-second foundation

3-SECOND CORE
Civil engineering enables buildings, roads, bridges and all the other structures in our built environment, ensuring that they are safe and durable enough to last for centuries.

3-MINUTE IDEA
Structural engineering, a part of civil engineering, is about ensuring that buildings, dams and bridges stand up. Calculations based on maths and physics principles predict how structures respond to the forces that nature throws at them – gravity, wind and earthquakes. Equally important, structural engineers predict forces during construction. That's why homes, schools and offices are safe to live in. Spectacular structures such as huge bridges reflect the work of structural engineers, as do invisible structures such as tunnels.

Civil engineering is about roads, railways, buildings, water, sewerage and much more. Civil engineers such as Eugène Belgrand and Joseph Bazalgette created sewer systems in Paris and London, eradicating cholera and saving the lives of millions. The Industrial Revolution reduced the cost of iron and steel in the nineteenth century and Thomas Telford showed how it could be used for bridges, canals and harbours. Together with Isambard Kingdom Brunel's railways, bridges and ships, these developments transformed transport. Today, civil engineers are working to solve challenges and improve people's lives – from creating flood defences and dams to building our largest infrastructures and tallest buildings. Every project poses challenges: there may be obstructions in the ground or poor soil conditions; tunnels or structures that the new construction needs to weave through; or constraints on finance and time. Engineers think creatively to work through these problems, try different solutions and choose the best option. Finding creative economic solutions for difficult construction challenges is immensely rewarding, and civil engineers embrace new technologies to create a world that can support future generations.

RELATED TOPICS
See also
GEOTECHNICAL ENGINEERING
page 40

ENGINEERS & ARCHITECTS
page 42

3-SECOND BIOGRAPHIES

SIR MARC ISAMBARD BRUNEL
1769–1849
French-born civil engineer, father of the more famous Isambard; created the 'Thames Tunnel', the first under a navigable river.

EMILY WARREN ROEBLING
1843–1903
Managed the Brooklyn Bridge construction for 11 years, after father-in-law, designer John Roebling, died; learned civil engineering from her incapacitated husband.

30-SECOND TEXT
Roma Agrawal

Civil engineers create the things we take for granted but would find hard to live without.

3 April 1929
Born near Dhaka, British India, now Bangladesh

1950
Graduates in Civil Engineering at Dhaka University; appointed assistant engineer, Highway Department

1952
Awarded Fulbright Scholarship and Pakistan Government Scholarship

1955
Completes PhD; employed by Skidmore, Owings and Merrill Architects (SOM), Chicago

1957
Director of Pakistan Building Research Centre, Karachi; Technical Advisor to Karachi Development Authority

1960
Returns to SOM, commences teaching at Illinois Institute of Technology

1963
43-storey DeWitt-Chestnut Apartment Building completed

1966
Appointed partner in Skidmore, Owings and Merrill

1967
Becomes naturalized American citizen

1969
John Hancock Center completed in Chicago, with tubular frame design

1971
Pioneers use of computers for structural calculations and design drawings

1981
Hajj Terminal at King Abdulaziz International Airport receives Agha Khan Award for Architecture

27 March 1982
Dies in Jeddah, Saudi Arabia

FAZLUR RAHMAN KHAN

Cities have been both the cradle and products of engineering for at least 8,000 years. Skyscrapers form the heart of modern cities because people can live and work close to each other, developing trusting relationships on which engineering, trade and commerce depend. The 'father of tubular designs', Fazlur Rahman Khan, transformed skyscraper design in the 1960s.

Born in Bengal, now Bangladesh, in 1929, Khan graduated in civil engineering from Dhaka University, winning a Fulbright scholarship to study in the US. After completing a PhD in 1955 researching reinforced concrete design, Khan went to work with Chicago architects Skidmore, Owings and Merrill, because they were happy to let him take charge of design and construction projects. The firm was renowned for skyscraper design. Khan soon realized that horizontal live loads from wind and earthquakes pose the greatest design challenges for tall building structures. He explored new ideas, working with students at the Illinois Institute of Technology and through public and professional lectures.

At the time, masonry shear walls between interior steel columns resisted horizontal loads. Buildings had to be rectangular, with little flexibility to change the internal layout. Khan's breakthrough was to design the outer shell of the building as a framed tube to resist horizontal loads, reducing the amount of steel needed by 40 per cent or more. His design eliminated almost all the interior columns and masonry walls, allowing unobstructed internal spaces. His buildings were taller and less expensive, and allowed architects to design almost any shape they wanted.

In 1963, the 43-storey DeWitt-Chestnut Apartment Building in Chicago was completed, the first skyscraper to use the structural tube frame design. The 110-storey Sears Tower, completed in 1973 and also in Chicago, was constructed as a bundle of parallel tube frames – as Khan described it, a group of pencils bundled together with a rubber band. Use of lightweight concrete and high-strength steel enabled buildings such as the 160-storey Burj Khalifa in Dubai – 828 m (2,700 ft) high. Experts consider that the ultimate height using the tube frame design has yet to be reached. Khan also pioneered innovative building forms such as cable-stayed roofs, notably for the immense Hajj Terminal at Jeddah Airport.

Khan was renowned not only for his technical expertise, but also for his humanity and love of art and literature. He epitomizes the US success narrative, a country that has consistently attracted and rewarded hard-working migrants. Khan died of a heart attack, aged 52, while on a trip to Saudi Arabia. His body was returned to the US and was duly buried in Chicago.

James Trevelyan

FORCE EQUILIBRIUM PRINCIPLE

30-second foundation

3-SECOND CORE
Engineers use the force equilibrium principle to predict forces in a structure, particularly the foundations.

3-MINUTE IDEA
Predicting storm, accident or earthquake loads accurately is almost impossible. Engineers rely on standards that prescribe the maximum loads that common structures must withstand. National standards tell engineers what wind loads to expect where, and how much weight each floor of a building must support. In situations not covered by standards, engineers rely on measurements or computer predictions.

The equilibrium principle enables engineers to analyse forces in a structure: it will only move if the combination of all forces acting on it is not zero. A free-body diagram helps identify the forces. For most structures, the main forces are static loads: gravity and reaction forces from the foundations. Live loads can be significant too: wind, earthquakes, moving vehicles, liquids sloshing in tanks, accidental impacts or machinery, even people walking. For coastal structures, wave forces are critical. Since forces act in different directions, engineers resolve them into components parallel to local x, y and z axes that define spatial reference directions at right angles to each other. The components can be positive or negative depending on their direction. Engineers add all the components together for each direction and then apply the force equilibrium principle. To remain stationary, the reaction forces from the structure's foundation must exactly balance the sum of all the applied loads. Engineers use these methods to calculate the design requirements for the foundations. Particularly in earthquakes or storms, one or more live loads may have to be resisted by tension forces in foundations that could, if large enough, result in them being pulled out of the ground.

RELATED TOPICS
See also
DIVIDE & CONQUER
page 14

CIVIL ENGINEERING
page 34

MECHANICAL ENGINEERING
page 56

3-SECOND BIOGRAPHIES

ROBERT HOOKE
1635–1703
Described the theory of elasticity, helping to invent many mechanisms.

ELMINA WILSON
1870–1918
First female civil engineer and professor in USA, helped design skyscrapers.

STEPAN PROKOPOVYCH TIMOSHENKO
1878–1972
The father of mechanics, the analysis of forces in machines and structures.

30-SECOND TEXT
James Trevelyan

Engineers analyse forces acting in different directions on structures.

+Z
–Y
–X
+Y
+X
–Z

GEOTECHNICAL ENGINEERING

30-second foundation

Soil and rock under the earth's surface are the domain of geotechnical engineering, a specialization for civil engineers. Everything we build needs foundations, and geotechnical engineers make sure that they are strong enough and that the soil underneath will not give way. Geotechnical engineers characterize soils by measurements of void ratio, density, water content and friction angle, the angle at which a soil surface begins to slide. Laboratory tests on bore hole samples provide data for design. As water content increases, especially with low porosity clay soils, bearing capacity decreases and larger foundations are needed for a given building. Drying can also cause soil shrinkage and cracking. Data from site surveys greatly influence cost estimates. Critical factors in rock include the location and direction of geological faults, historical cracking of the earth's crust. Vertical faults can be missed easily when surveying a site with drill holes and can cause water leakage under dam walls. Detecting faults is critical when assessing the stability and safety of mines and tunnel construction. Geotechnical engineers will be onsite during construction to check that earth works behave as expected, making adjustments when needed. Earthquakes can pose acute challenges, as some soils liquefy when shaken after heavy rain.

3-SECOND CORE

All man-made structures need foundations to remain standing. Geotechnical engineers investigate the soil and rock below to design foundations that may need to last for centuries.

3-MINUTE IDEA

Geotechnical engineers play an important role in mining. They advise on pit slope stability, shaft and tunnel excavation and waste storage. They design tailings dams to contain waste pumped from the processing plant as a slurry – crushed minerals carried by flowing water. The water is recovered and returned to the plant. Engineers find economic solutions with hydrogeologists and environmental engineers to ensure the waste is safely contained for centuries with minimal environmental impact.

RELATED TOPICS

See also
CIVIL ENGINEERING
page 34

TAMING GREAT RIVERS
page 46

ENGINEERING ETHICS
page 48

3-SECOND BIOGRAPHIES

HENRY DARCY
1803–58
Developed 'Darcy's Law', defining the flow of water through a porous medium.

ALBERT ATTERBERG
1846–1916
Established 'Atterberg limits', which help distinguish silt from clay and provide guidance for geotechnical engineers.

KARL VON TERZAGHI
1883–1963
Established the fundamentals of geomechanics; developed the effective stress principle.

30-SECOND TEXT

Doug Cooper

Engineers decide how much material can be removed safely, without triggering a mine collapse.

ENGINEERS & ARCHITECTS

30-second foundation

3-SECOND CORE
Engineers and architects bring complementary skills and knowledge to create great buildings, collaborating closely from start to finish.

3-MINUTE IDEA
Engineers and architects collaborate from the start on major buildings, helping clients and regulators understand conceptual possibilities. Designs evolve within regulatory, economic, structural and subsoil limitations. Construction practicalities often dictate extensive redesign, negotiation and ingenuity to preserve agreed visual appearances. Engineers provide the close supervision needed to ensure safety for everyone, and to retain sufficient alignment with design intent while accommodating the interests of everyone involved.

Engineers and architects have complementary and mutually dependent roles in creating buildings. Architects focus on visible aspects: the exterior, interior layout, finishes and ambience. Engineers work almost entirely inside the hidden spaces behind the walls, floors and ceilings. They design foundations, structure and services such as air conditioning, lighting, security, communications, water, gas, electricity and sewerage, and renegotiate visible boundaries when hidden spaces are insufficient. While architects may seek accolades for visual appearance, engineers are legally responsible for the structure and safety, during construction and for decades after completion. Architects hone their skills through years of studio workshops; engineers use abstract concepts to analyse and predict stress fields and invisible airflow through building spaces. Conceptual plans can retain some fluidity in the mind of an architect. Engineers, mindful of their hourly fees and prior estimates, prefer a fixed scope of work. Architects, commissioned by the owner, receive an agreed percentage of the building cost. Engineers provide their services to the architect or project manager. Unsurprisingly, with such different backgrounds, thinking and financial interests, close cooperation requires hard work and careful listening.

RELATED TOPICS
See also
CIVIL ENGINEERING
page 34

GEOTECHNICAL ENGINEERING
page 40

ENVIRONMENTAL ENGINEERING
page 50

3-SECOND BIOGRAPHIES

SÉBASTIEN VAUBAN
1633–1707
Responsible for defensive fortifications for several hundred French cities.

OVE NYQUIST ARUP
1895–1988
Translated architect Jørn Utzon's ideas into structures for the Sydney Opera House.

SANTIAGO CALATRAVA
1951–
Engineer, artist and architect: a rare combination resulting in visually stunning buildings.

30-SECOND TEXT
James Trevelyan

A great building is testament to a successful collaboration between engineer and architect.

ENGINEERING & CIVILIZATION

30-second foundation

Engineering has enabled civilization, supporting humans to move from being hunter-gatherers to being inhabitants of towns and cities, by building roads, bridges and aqueducts. Water was diverted from streams to facilitate sustainable village agriculture; systematic irrigation along the Nile enabled the prosperity of Egypt and its civil engineering legacy – pyramids and enormous temples. Similarly, Mohenjo-daro in Pakistan dates from 2500 BCE and is one of the world's earliest major cities. Located in the Indus Valley, it drew water from groundwater wells and channelled away wastewater. It was laid out on a systematic grid like many modern cities. Roman engineers later mastered cement using volcanic ash, constructing dams, aqueducts, vast urban water supply systems and sewers, promoting vital public health for one million people. Other extraordinary engineering achievements include Persepolis in Iran and the abandoned city of El Mirador in Guatemala, which dates from the sixth century BCE and contains many pyramids, one of them amongst the largest in the world by volume. In China, many grand cities were built from as early as 3000 BCE. China has also contributed four great inventions: the compass, gunpowder, papermaking and printing, along with many others.

3-SECOND CORE

Ancient civil engineers made cities, linked by roads, some of which are still in existence. They brought clean water, removed waste and built stunning structures.

3-MINUTE IDEA

'Engineer' first emerged in medieval English: a person who operated engines, likely from the Latin word ingenium, meaning clever, ingenious. Creations such as the Egyptian and Central American pyramids, Persepolis in Iran and the Acropolis at Athens required people with extraordinary abilities. The Roman engineer Vitruvius later described the need for technical foresight, planning and coordination of many skilled workers. These people, the earliest engineers, gained the confidence of rulers, who provided the immense resources required.

RELATED TOPICS

See also
CIVIL ENGINEERING
page 34

GEOTECHNICAL ENGINEERING
page 40

ENGINEERS & ARCHITECTS
page 42

3-SECOND BIOGRAPHIES

MARCUS VITRUVIUS POLLIO
ca. 80–15 BCE
Created bridges, buildings and aqueducts for Rome.

SEXTUS JULIUS FRONTINUS
ca. 40–103 CE
Documented the nine aqueducts of Rome, reduced water theft and improved maintenance.

APOLLODORUS OF DAMASCUS
second century CE
Led the construction of Trajan's Bridge over the Danube.

30-SECOND TEXT

Roger Hadgraft

The Pantheon is still the world's largest unreinforced dome, built from lightweight concrete.

TAMING GREAT RIVERS

30-second foundation

The development of high-strength steel and steel-reinforced concrete in the early twentieth century transformed civil engineering. Concrete is weak in tension; embedded steel resists tension loads. Empowered with these new materials, engineers have tamed all the world's major rivers in just a few decades. Bridges have enabled roads and railways to cross rivers and gorges that separated populations for millennia. Dams have brought floods under human control, enabling vast irrigation schemes that have transformed deserts into productive agricultural land, hugely increasing food production. Dams also provide essential reliable water supplies for most of the world's cities and industrial centres. Turbines under the dams transform the energy of flowing water into electric power, by far the largest source of renewable energy. Many challenges remain, however. Silt from erosion, particularly from over-used agricultural land, accumulates in reservoirs behind dams, significantly decreasing water storage and hydro-electricity production. Water logging and salt accumulation have eroded agricultural production in many irrigation projects. Governments anxious to demonstrate rapid progress by building dams and bridges have been reluctant to provide sufficient maintenance resources.

3-SECOND CORE
Steel and reinforced concrete enabled engineers to build bridges and dams, taming the world's great rivers in the twentieth century to provide water and energy, and eliminate rivers as transport obstacles.

3-MINUTE IDEA
Huge construction projects fostered the growth of huge engineering enterprises, both government and privately owned. Their elaborate organizational processes successfully choreograph technical collaboration across vast engineering teams and with clients and suppliers. While technical collaboration is a central aspect of engineering practice, methods that enable successful large-scale technical collaboration are seldom recognized as core engineering knowledge.

RELATED TOPICS
See also
CIVIL ENGINEERING
page 34

GEOTECHNICAL ENGINEERING
page 40

ENGINEERS & ARCHITECTS
page 42

3-SECOND BIOGRAPHIES

ERNEST LESLIE RANSOME
1852–1917
Pioneered modern techniques for using reinforced concrete.

MOKSHAGUNDAM VISVESVARAYA
1861–1962
Led development of flood protection systems and major water storage dams in India.

ZHENG SHOUREN & ZHANG CHAORAN
1940–
Led the design and construction of the Three Gorges Dam.

30-SECOND TEXT
James Trevelyan

Steel and reinforced concrete enabled construction of dams and pipelines.

ENGINEERING ETHICS

30-second foundation

A piling contractor building the foundations for two tower blocks does not drive the piles down to bed-rock as per the design. Instead, to save money, the contractor only drives the piles part way, and deceives the government that the works have been built correctly. When the buildings reach 34 floors in height, they begin to tilt. Both buildings are demolished, the contractor goes into liquidation and three people are jailed for fraud. Luckily, no one dies. In other cases, buildings have collapsed due to circumvention of building standards – bribes were paid to inspectors to overlook defects. Quite simply, corruption in engineering kills. Ethics in engineering is vital to ensure good quality and safe infrastructure at fair value. To be ethical in engineering means: not to pay or receive bribes (in return for a contract); not to commit fraud (covering up defects); not to participate in a cartel; to avoid conflicts of interest (awarding a contract to family members); to provide honest and impartial advice. Organizations can promote good ethics by strong leadership, training, implementing management controls and encouraging the reporting of bad practice. Disciplinary procedures can help to enforce compliance.

3-SECOND CORE
Engineering and ethics are intertwined. Engineers take decisions that affect other people, or people yet to be born, and have to be mindful of the consequences to achieve the best possible results.

3-MINUTE IDEA
Previously, prosecutors rarely bothered with bribery, fraud or cartels in engineering. However, realization of the damage they cause has led to numerous recent prosecutions. Many major organizations have been fined for bribery, fraud or cartels – and some managers have been jailed. Such malpractice has a substantial negative effect on an organization's finances and reputation. The good news is that the companies in question have since implemented ethical programmes to prevent repetition.

RELATED TOPICS
See also
GEOTECHNICAL ENGINEERING
page 40

THINKING DIFFERENTLY
page 134

3-SECOND BIOGRAPHIES

THEODORE COOPER
1839–1919
American civil engineer and chief designer for the first Quebec bridge which collapsed while under construction in 1907, with 75 fatalities; criticized in post-accident reports for his judgement.

ROGER MARK BOISJOLY
1938–2012
American mechanical engineer who strenuously opposed launching space shuttle Challenger in January 1986, citing a likely failure of booster O-rings. NASA ignored the warnings.

30-SECOND TEXT
Neill Stansbury

Engineering institutions develop ethical codes to guide practice and decision-making.

ENVIRONMENTAL ENGINEERING

30-second foundation

Environmental engineers assess an engineering project in terms of energy efficiency; effects on air, water, land, flora and fauna; human health risks; noise emissions; conservation measures; and use of natural resources. They influence engineering designs, construction plans, process operations and project economics to maximize environmental protection. Environmental engineers analyse a project as part of a closed system on the earth, drawing on multidisciplinary perspectives and critical thinking skills. They use computer models of material and energy flows, the atmosphere and climate, the water cycle and the carbon cycle – all to predict how pollution will disperse in soil, waterways and the atmosphere. These models are built using equations that represent physical, chemical and biochemical principles; for example, conservation of mass and energy. Social impact assessments draw on sustainable development principles and social science research methods such as focus groups. Environmental engineers work on urban design projects and often use their knowledge to find clever ways to improve urban environments, such as turning water drains into elaborate parks and nature reserves, while providing water storage to reduce flows during peak flood events.

3-SECOND CORE

Environmental engineers apply engineering and scientific principles to develop solutions for pollution and to prevent damaging effects on the global environment.

3-MINUTE IDEA

Environmental Impact Assessment (EIA) examines the consequences of a project, both positive and negative, covering all aspects of a project, including planning, construction, operation and end of life. This process is governed by legislation in most countries and includes input from the public. The outcome of this process requires decision makers to account for environmental values in their decisions, taking into account detailed environmental studies and potential impacts for current and future generations.

RELATED TOPICS

See also
THINKING DIFFERENTLY
page 134

RESOURCE SCARCITY
page 142

CONTROLLING POLLUTION
page 148

3-SECOND BIOGRAPHIES

ELLEN SWALLOW RICHARDS
1842–1911
American civil engineer and the first female admitted to Massachusetts Institute of Technology, Richards pioneered work in sanitary engineering, water supply and public health.

RACHEL CARSON
1907–64
American marine biologist, author and conservationist who attracted world attention by highlighting the impact of pesticides on the environment and started the global environmental movement.

30-SECOND TEXT

Julia Lamborn

Environmental engineers create sustainable solutions to help people and our planet.

MECHANICAL, MATERIALS & MECHATRONIC ENGINEERING

MECHANICAL, MATERIALS & MECHATRONIC ENGINEERING
GLOSSARY

actuator Machine component that moves or controls another component, such as an electrically operated valve regulating fluid flow moving a piston; a cylinder operating an excavator loading arm.

alloy Mixture of metals and other elements to improve material properties. Soft aluminium becomes strong like steel when alloyed with zinc, magnesium, copper and other elements.

atomic structure Arrangement of atoms in a solid; regular, as in crystals and most metals, or irregular, as in glass, or a combination, as in many ceramics.

availability Proportion of time that machinery operates with specified performance.

corrosion Progressive spoiling of exposed metal surfaces caused by chemical transformation to oxides or other by-products, for example, rusting of steel.

ceramic Hard, usually brittle, non-metallic materials, often used in high temperature applications or where excellent insulation is needed.

composite material Material comprising two or more distinct materials with complementary properties. Reinforced concrete consists of cement, stones and steel – the steel enables brittle cement and stones to withstand tension stress.

energy conversion, transformation Energy can be converted from one form to another; for example, an electric motor converts electric energy into rotational mechanical energy; a battery converts chemical energy into electric energy.

energy intensity Amount of energy needed to produce a given quantity of material or to achieve a given result.

fatigue Progressive failure of metal components subjected to repeated cyclical load, for example, loaded axles. Engineers have to allow for a lower maximum stress to avoid fatigue failure in these components.

feedback control System that automatically regulates the state of a machine by measuring the state of the machine with sensors and feeding back the measurements to adjust actuator settings, to compensate for external disturbances.

heat treatment Using heating and cooling to change material properties; for example, if steel is heated and rapidly cooled in water, it becomes much harder and brittle.

laminar flow Smooth fluid flow with no turbulence, typical of slow-speed flow.

kinetic energy Energy associated with a body or fluid that is moving.

polymer material Material consisting of molecules which are long chains of simpler molecules. For example, polythene plastic has long chain molecules of ethylene, hence 'poly-ethylene', shortened to 'polythene'. Some polymers like Kevlar form extremely strong materials.

potential energy Stored energy – for example in a compressed spring, a battery, or water in an elevated reservoir – that can be released at a later time. A pendulum illustrates the regular exchange between potential energy when it is stationary at the ends of the swing, and kinetic energy when it is moving.

reliability Time performance of machinery or plant with no failures, usually measured as mean time between failures (MTBF).

rolling (material processing) Metal properties can be improved by squeezing the metal between high-pressure rollers, making it thinner, longer and stronger.

sensor Device that measures a physical property and generates a signal indicating the measured value. A thermocouple measures temperature and generates a small electric voltage indicating the temperature.

sustainment (engineering asset management) Combination of operating practices, planned maintenance, repairs and pre-planned refurbishment or replacement to maximize the proportion of time that equipment performs as expected.

thermodynamics Physical principles governing the movement of heat and energy transformations that guide engineers working with engines, air conditioning and chemical processes.

turbulent flow Erratic fluid flow with eddies and rapid small random variations, typical of high-speed fluid flow.

trade-off Compromise between two or more desirable but incompatible results requiring human judgement to decide.

MECHANICAL ENGINEERING

30-second foundation

Mechanical engineering guides design and manufacture for moving objects and fluids in motion – from machines, tools and engines, to oil platforms and even artificial hearts and blood vessels. Newton's three laws of motion provide fundamental principles used by mechanical engineers to explain how machines work. Mechanical engineers became important during the Industrial Revolution from 1750 onwards, applying Newton's principles and creating pumps, machine tools, spinning machines, railways, ships and, later, automobiles. Bernoulli's principles then extended Newton's laws to moving fluids, enabling engineers to design pumps and piping systems, providing drinking water and removing sewerage waste, enormously improving health. The laws of thermodynamics explained the release of chemical energy in heat engines, enabling mechanical engineers to create faster and more efficient automobiles, aircraft and spaceships, as well as energy-efficient air conditioning. Machine tool developments started the progressive automation of factories, providing high quality manufactured goods at ever-decreasing prices. Improved materials combined with powerful computer analysis and 3D printing are greatly increasing design possibilities.

3-SECOND CORE

Mechanical engineering guides design for moving objects and fluids, particularly cars, aircraft and water supply systems. Mechanical engineering enables machine tools and automation in factories.

3-MINUTE IDEA

The Carnot Cycle, named after the French engineer, describes an ideal heat engine and explains why we cannot extract all the energy in fossil fuels and convert it to electricity. It also explains how we can use mechanical energy for cooling, reversing the natural flow of heat towards cooler materials. Improving energy efficiency by improved mechanical design is one of the easiest ways to minimize greenhouse gas emissions and reduce global warming.

RELATED TOPICS

See also
MATERIALS ENGINEERING
page 60

MECHATRONICS
page 62

DEFENCE READINESS
page 64

3-SECOND BIOGRAPHIES

DANIEL BERNOULLI
1700–82
Swiss mathematician who described how the kinetic energy of a moving fluid corresponds to potential energy of an equivalent depth of fluid at rest.

NICOLAS LÉONARD SADI CARNOT
1796–1832
French mechanical engineer who explained why heat engine power is proportional to the temperature difference between the hottest and coolest parts.

30-SECOND TEXT

James Trevelyan

Movement of machines, mechanisms and fluids is the core issue in mechanical engineering.

L
$\ln\left(\frac{T_F'^2}{T_{i1} \cdot T_{i2}}\right) = 0$

5 July 1820
Born in Edinburgh, Scotland

1836–38
Edinburgh University: wins awards for essays 'Undulatory Theory of Light' and 'Methods of Physical Investigation'; graduates in Civil Engineering

1839
Employed by Sir John MacNeill

1842
Elected fellow of Royal Scottish Society of the Arts

1843
Publishes 'Fracture of Axles', in which he identifies metal fatigue failures

1844
Employed by Locke and Errington

1848
Commences research on molecular physics and thermodynamics

1849
Elected fellow of the Royal Society, Edinburgh

1852
Designs water supply for Glasgow from Loch Katrine

1854
Awarded Royal Society of Edinburgh Keith medal for research on thermodynamics; appointed Regius Professor of Civil Engineering and Mechanics, University of Glasgow

1862
Publishes *Manual of Civil Engineering*

1863
Awarded Gold Medal of Institution of Engineers in Scotland for paper on 'Liquefaction of Steam'

1866
Publishes 'Shipbuilding – Theoretical and Practical'

1868
Elected to Royal Academy of Sweden

1869
Publishes 'Machinery and Millwork'

1872
Investigates flour mill explosions, reports on causes

24 December 1872
Dies in Glasgow

WILLIAM JOHN MACQUORN RANKINE

Few mechanical engineering topics have had more impact on all our lives than heat engines. Engines transform heat energy – solar, nuclear, fossil and geothermal – into usable mechanical and electrical energy. Scottish engineer Macquorn Rankine provided the clear theoretical and practical understanding engineers needed to perfect the automobile engines, generator turbines jet and rocket engines we use today.

Born in Edinburgh in 1820, ill-health in childhood kept Rankine from school. His father and private tutors provided much of his education, until an uncle presented him with Newton's *Principia*, from which he taught himself advanced mathematics and mechanics. He studied chemistry and civil engineering at Edinburgh University, and then worked as a civil engineer for 16 years before joining the University of Glasgow as Professor of Civil Engineering and Mechanics.

The sheer volume of technical and professional articles, books and papers Rankine published is staggering: several hundred between 1842 and 1872. However, even more amazing is the clarity and simplicity of his writing, which has enabled so many engineers to build on his ideas. Much of what he wrote forms the standard texts that mechanical engineers study today around the world. Most were original scientific contributions, based on self-taught knowledge.

Rankine collaborated with Rudolf Clausius and William Thomson (Lord Kelvin), after whom the absolute temperature scale is named. Together they refined French engineer Sadi Carnot's formulations in his book *Motive Power of Heat*, published two decades earlier. They formulated the theoretical understanding on the equivalence of heat and mechanical energy, and the transformation between the two, which explains how heat engines perform useful work.

The breadth of Rankine's understanding is truly astounding: apart from the theory of heat engines, his contributions included practical methods to plan railway tracks, research on liquid-to-gas and gas-to liquid transitions, simple and practical methods to predict the power required to drive steam ships based on confidential data provided to him by commercial ship builders, and the shape and effects of wind waves on ship stability.

Rankine was much admired by colleagues, not only for technical contributions: 'His unfailing amiability of temper, the generosity of his mind and the warmth of his affections, made him as dear in the circle of his friends as he was distinguished in the world of science.' He died on Christmas Eve, 1872, aged only 52. He was unmarried and had no children.

James Trevelyan

MATERIALS ENGINEERING

30-second foundation

Materials are chosen for their unique properties: strength, flexibility, corrosion-resistance, electrical conductivity, magnetism and even colour and surface appearance. There are four types of materials: metals, ceramics, polymers and composites. Metals such as steel, aluminium and copper are the most widely used due to strength and flexibility. Ceramics are hard and brittle with excellent corrosion and heat resistance. Plastics are usually softer, pliable and easily moulded. Composites combine different materials to obtain special properties, and include glass and carbon-fibre reinforced plastics, bone, concrete and timber. Materials engineers use 'processing-structure-property' principles. Material composition, how it is made and subsequent processing all influence atomic structure, determining properties. Materials engineers find ways to process materials to obtain desired properties. For example, alloying, heat treatment and rolling make metals stronger. Technological advances often rely on new materials: the Space Shuttle required special ceramic tiles to protect it from intense heat; energy-saving lights rely on materials that emit intense light when electric current flows through them; refrigeration and air conditioning rely on special fluids.

3-SECOND CORE

Materials engineering guides design, modification and selection of materials used by engineers. Materials engineers create new materials for special applications by modifying composition and atomic structure.

3-MINUTE IDEA

Hooke's Law, named after Robert Hooke, a seventeenth-century British polymath, states that a material deforms in proportion to applied stress, up to a limit, and returns to its original shape when released. The ratio of stiffness to deformation is a key material property. A stiff material like concrete will not deflect much, even under very heavy loads. A plastic drinking straw bends easily, even with a light force.

RELATED TOPICS

See also
MECHANICAL ENGINEERING
page 56

NANOTECHNOLOGY
page 104

AEROSPACE MATERIALS
page 124

3-SECOND BIOGRAPHIES

ADOLF MARTENS
1850–1914
Discovered how heating and cooling steel produced a different grain structure.

ALFRED WILM
1869–1937
Developed high-strength aluminium alloys that enabled use in aircraft.

STEPHANIE KWOLEK
1923–2014
Invented Kevlar, extremely high-strength polymer fibres used in bullet-proof vests.

30-SECOND TEXT

Tim Sercombe

Materials engineering lays the foundations for all other disciplines of engineering.

MECHATRONICS

30-second foundation

3-SECOND CORE

Mechatronics describes systems with mechanical, electrical and electronic devices working together, usually with computers. These systems help to make complex machines safe and extremely reliable.

3-MINUTE IDEA

Spare a thought for the engineers who design and maintain these complex control systems. Designers aim for reliable and safe operation under normal conditions. A design that allows sensors and other parts to be disconnected for repairs and yet still provides complete, constant safety is much more difficult. Software bugs can remain undetected because maintenance is infrequent. Maintenance technicians need to distinguish persistent software faults from random component failures.

By the 1980s, electronics and micro-computers controlled machines such as robots and car engines. Companies soon needed specialist engineers to design electric machines with sensors and connect them with computers. These engineers also wrote software, because they understood details of the machines. First in Japan, later elsewhere, these engineers became known as mechatronic engineers. Mechatronics has enabled the age of smart machines that can adapt their behaviour. A computer ensures that car engines start easily and run smoothly even in the coldest weather and use the least fuel while minimizing exhaust emissions; another computer senses the key approaching and automatically unlocks the door. Feedback control is one of the main principles used in mechatronics. A car's cruise controller adjusts engine power by feeding the speed sensor signal back to the controller: when the car slows, more fuel is supplied, boosting power; if the speed is too fast, it reduces engine power. Many cars now can automatically follow the car in front on a motorway using radar sensors to measure the distance, with automatic braking if needed. Safety and reliability are the main mechatronics challenges. Systems such as anti-lock brakes (ABS) are so reliable that they can be depended on for safety.

RELATED TOPICS

See also
ROBOTICS & AUTOMATION
page 70

COMPUTER ENGINEERING
page 98

DRIVERLESS CARS
page 128

3-SECOND BIOGRAPHIES

ÁNYOS ISTVÁN JEDLIK
1800–95
Invented practical electric motors in 1828.

ERNST WERNER SIEMENS
1816–92
Developed electric telegraphs and motors, and founded the Siemens company.

ROBERT BOSCH
1861–1942
Created reliable spark plugs for automobiles, introduced eight-hour work days and ensured his company profits benefitted charities.

30-SECOND TEXT

James Trevelyan

Smart machines sense impending faults and alert technicians when your car is serviced.

tmpFormat
str(key))
tempString
tmpFormat))))
"BUFFER"):
elif(typeOf
replace(
"ASCII
tempString

DEFENCE READINESS

30-second foundation

Defence investment is seen as an insurance policy against potential military actions by others. Defence engineers have to ensure that expensive military equipment will work when needed, maybe 30–40 years after acquisition, while minimizing sustainment costs. Rapid advances in electronics and computer technology have imposed further challenges. Original electronics almost certainly have obsolete components, so large quantities of spares have to be stored. Aircraft, ships and vehicles may require complete and expensive replacement of their electronics just to maintain reasonable capabilities, taking them out of service for months, even years. High quality manufacturing methods have evolved to meet these challenges, and to provide safety for civilian aircraft. Traceability is one such manufacturing method: recording each batch of material, each production run, even each worker's contributions in making components. Engineers predict reliability and availability, and use statistics to decide which improvements could be most effective. Detailed observation and record keeping is essential, as are highly trained maintenance technicians.

3-SECOND CORE

The need for engineers to maintain the readiness of expensive defence equipment has led to high quality manufacturing methods, now used to make highly reliable consumer products.

3-MINUTE IDEA

In the mid-twentieth century, there were automotive or electrical repair shops at practically every street corner. Breakdowns were a regular and accepted part of life. Today, car-makers provide 100,000-km (60,000-mile) warranties, and those repair shops have disappeared. High quality manufacturing methods developed for defence and civilian aviation have now become almost universal. We all benefit from countless small improvements made by thousands of engineers in factories everywhere.

RELATED TOPICS

See also
MECHANICAL ENGINEERING
page 56

MATERIALS ENGINEERING
page 60

AEROSPACE MATERIALS
page 124

3-SECOND BIOGRAPHIES

KONOSUKE MATSUSHITA
1894–1989
Founded Panasonic, one of the earliest companies to adopt Deming's ideas on quality manufacturing.

W. EDWARDS DEMING
1900–93
Influential in the development of high quality manufacturing.

GEOFFREY LIGHT WILDE
1917–2007
Led developments in highly reliable aircraft engines at Rolls Royce, which are used in most modern civilian aircraft.

30-SECOND TEXT

James Trevelyan

Engineers predict how many hours an aircraft or ship will operate for without maintenance.

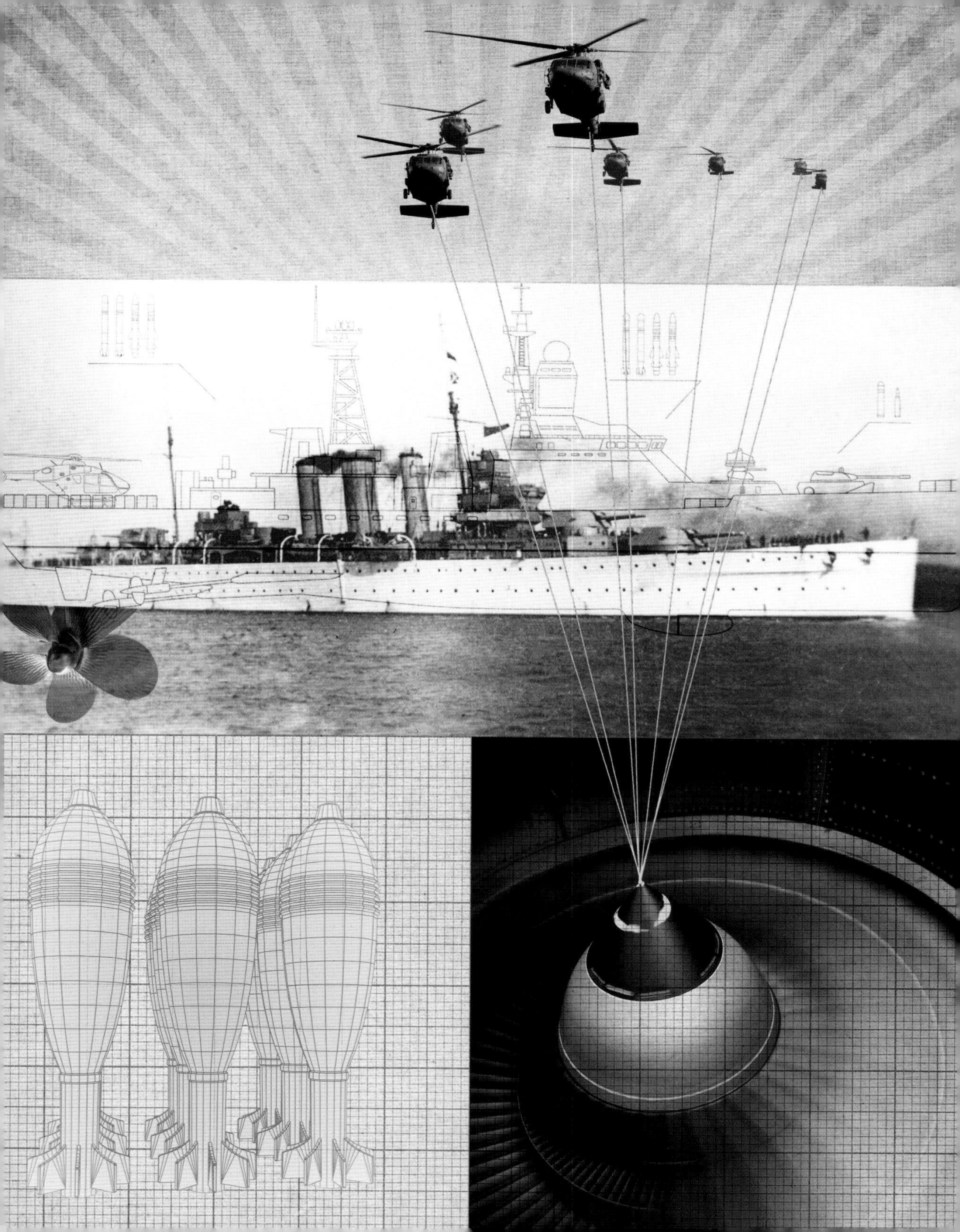

THRUST BEARINGS TO TERABYTES

30-second foundation

3-SECOND CORE

Mechanical engineers invented the tilting pad thrust bearing to transfer propeller thrust to the ship's hull. The same principle enables hard disk recording heads to fly nanometres above the disk surface.

3-MINUTE IDEA

Disk heads fly over the spinning surface on a thin film of air. Force from a flexible arm balances the lifting force from the thin film of gas. But what if the power goes off? What stops the head rubbing on the disk when it stops? The control computer senses the power loss and quickly moves the head over the innermost track before the power fails completely. When the disk stops, the head comes to rest on an unused section of the disk.

The invention of propellers for ships required a second great invention for their true potential to be realized. Propellers rotate, but the ship does not. A special thrust bearing is needed to transfer the propeller force from the shaft to the ship's hull. With large ships, the force can be 200 tonnes or more. In 1905, two mechanical engineers simultaneously conceived the same idea independently. Anthony Michell in Australia and Albert Kingsbury in America invented the tilting pad thrust bearing. A slightly tilted steel plate immersed in oil will skate over a rotating disk. Relative movement squeezes oil between them and keeps them apart. Six plates in a 150-cm (60-in) circle are sufficient for the largest ships. Hard disk designers faced a different challenge – how to keep a magnetic recording head at 10,000 nanometres above a spinning disk surface (a human hair diameter is about 90,000 nanometres). The designers used the same solution: a slightly tilted head flies over the disk surface on a thin film of inert gas inside the housing. In the early 1970s, hard disk drives came with 5 megabytes on 40-cm (16-in) diameter disks. Forty years later, the recording head height had reduced to just 5 nanometres, and pocket-sized disks can store a million times more data.

RELATED TOPICS

See also
MECHANICAL ENGINEERING
page 56

COMPUTER ENGINEERING
page 98

FLOATING FACTORIES
page 120

3-SECOND BIOGRAPHIES

OSBORNE REYNOLDS
1842–1912
Irish mathematician, inventor and engineer who studied laminar and turbulent flow in fluids. The Reynolds number tells engineers when laminar flow becomes turbulent.

ALBERT KINGSBURY
& ANTHONY MICHELL
1863–1943 & 1870–1959
American and Australian mechanical engineers who simultaneously invented the tilting pad thrust bearing, now used for ship propeller shafts.

30-SECOND TEXT

James Trevelyan

Tilting pad thrust bearings revolutionized ships and hard disk drives.

WIND ENERGY

30-second foundation

Wind-powered boats traversed the Nile 2,000 years ago; Chinese farmers used wind-powered water pumps; ancient Persians milled grain using wind turbines with woven reed sails. Modern wind turbine blades are designed like aeroplane wings to harvest as much of the wind's energy as possible. A generator transforms shaft rotation into electricity. Traditionally, most wind turbines have a 'horizontal axis', meaning that the sails rotate around a line parallel to the ground, and the axis must be aligned with the wind. Carbon-fibre materials enable engineers to design lightweight 80-m (260-ft) long blades. Computer-controlled machines produce special blade profiles designed to reduce noise. With each rotor producing more power, specialized cooling systems are needed to protect both the electronics that dynamically adjust the blade and axis orientation to get the most power possible, and the gearboxes that step up axis rotation to the higher speeds of electricity generation. 'Vertical axis wind turbines' resemble the beaters of a mixer, and don't need to be aligned with the wind – though they do require a jump start. Wind strength varies, so complementary power sources are needed to balance electricity demand with supply.

3-SECOND CORE

Wind farms, often offshore, generate electricity as cheaply as fossil fuels. Impacts of wind power – social, environmental, economic and technological – involve trade-offs, and can still be controversial.

3-MINUTE IDEA

Wind results from the solar heating of our atmosphere and provides clean energy, although there are trade-offs. Carbon-fibre production is more energy-intensive than metals, though it can last longer. Natural composites are greener but not yet as easy to work with as traditional engineering materials. Wind turbines are noisy and can be hazardous to birds. Engineers choose designs that achieve economic energy production with acceptable social and environmental impacts.

RELATED TOPICS

See also

POWER GENERATION & ENERGY STORAGE
page 78

ELECTRICAL ENGINEERING
page 94

AEROSPACE MATERIALS
page 124

3-SECOND BIOGRAPHIES

HERO OF ALEXANDRIA
ca. 10–70 CE
Invented a wind-powered musical organ, a holy water vending machine and pumps.

JAMES BLYTH
1839–1906
Produced the first electricity-generating wind turbines using a vertical axis design.

GEORGES DARRIEUS
1888–1979
Developed the modern vertical axis wind turbine design.

30-SECOND TEXT

Jenn Stroud Rossmann

Harnessing renewable wind energy is an old idea that engineers continue to improve.

f_1
a
IV
g
a
f_2

ROBOTICS & AUTOMATION

30-second foundation

3-SECOND CORE

Most robotics and automation engineers aim for productivity and quality improvements in factories and warehouses. Robots perform best in specially designed environments that help them know where they are.

3-MINUTE IDEA

Artificial intelligence has long been seen as the breakthrough that would confer 'dumb' robots with intelligence, enabling them to work in unstructured environments, including disaster relief. However, even with 'deep learning', a promising recent development, this is still a distant vision. Autonomous drones and vehicles are evolving with cheaper, more powerful and reliable sensors, motors and batteries, and virtual reality will help people control robots more easily than before.

Robots are the most recent development in tools that extend human capabilities. The ones that preoccupy most robotics engineers are far from mechanical humanoids empowered with artificial eyes and the machine intelligence of science fiction. Most factory automation uses machines for specific tasks, such as forming a moulded plastic electric plug, complete with wires. Robots are preferred when complex movements are needed. Creating robots requires extensive collaboration: mechanical structure and mechanism design; gearing and brakes; electrical motors, batteries, hydraulics, electronic sensors, data communication and computers. Engineers push ingenuity to the extreme limits of material economy, strength and durability. Software development usually absorbs the greatest effort. Factory robots repeat the same movements precisely with minor adaptation. They need an entire production facility around them, with conveyors to bring parts, clamps to hold them and tools such as spot-welding guns. Safety interlocks enable technicians to safely repair breakdowns or make changes. Designing, building, programming and testing the production facility takes the most effort. It must recover quickly from faults, such as a malformed part jammed in a clamp.

RELATED TOPICS

See also
MECHATRONICS
page 62

DRIVERLESS CARS
page 128

FUTURE TRANSPORT: DRONE SHIPS
page 150

3-SECOND BIOGRAPHIES

KAREL ČAPEK
1890–1938
Playwright who introduced the term 'robot' for a human-like worker.

GEORGE C. DEVOL JR
1912–2011
Filed US Patent 2,988,237 for a 'programmed article transfer' machine, later called Unimate.

JOSEPH ENGELBERGER
1925–2015
First commercialized industrial robots for dangerous industrial work, and was an evangelist for robotics.

30-SECOND TEXT

James Trevelyan

Mobile robots are appearing in ports, mines, warehouses and even hospitals.

CHEMICAL ENGINEERING & ENERGY PRODUCTION

CHEMICAL ENGINEERING & ENERGY PRODUCTION

GLOSSARY

base load power Amount of electricity generation that is required continuously, 24 hours a day, to meet a variable demand. Additional 'peak load' power is needed as demand rises from the 'base load' level.

demand (electrical engineering) Requirement for electricity to be provided, determined by the number and power requirements of all devices currently connected and switched on in an electricity supply network.

effluent Liquid flowing away from a process plant, either discharged into a sewer, river or ocean or conveyed to another process plant for further processing.

electrolyser Reaction vessel or cell in which electricity drives a chemical transformation; for example, converting water into hydrogen and oxygen. A fuel cell does the opposite.

fuel cell Reaction vessel or cell in which gases and liquids combine and generate electricity; the opposite of an electrolyser.

general arrangement drawing (GA) Shows all the main components of a process plant or other machinery, explaining how they are arranged relative to each other in three-dimensional space; usually drawn to scale. Detailed drawings show the physical details of individual components needed for manufacturing and assembly.

geothermal energy Energy derived from hot rock formations deep in the earth's crust.

grid (electrical engineering) Entire electric power supply system, including electricity generating stations, transformers, switch yards and powerline interconnections.

inertia (spinning) Tendency of a spinning wheel to continue spinning – in effect, its rotational kinetic energy.

mass and energy balance Accounting for all the material and energy inputs and outputs in a single process unit using chemical reaction equations and equations of state that predict physical and chemical transformations.

nuclear fission Splitting of nucleii of a large heavy atom, such as uranium, after absorbing neutrons. Typically the mass of by-products is slightly less, and the remaining mass is transformed into a large amount of energy.

piping and instrumentation diagram (P&ID) Shows all the tanks, pumps, reaction vessels, piping connections and instrumentation needed to create a process plant and regulate its operation.

power reactor (nuclear engineering) Shielded vessel containing nuclear fission fuel and control devices to enable regulated production of heat energy.

process flow diagram (PFD) Shows all the process units incorporated in a process plant; used to explain how the incoming solids, liquids and gases are combined and transformed into the products.

process plant (chemical, mechanical, mining engineering) Substantial installation comprising tanks, reaction vessels, pumps, pipework and associated machinery for combining and transforming solids, liquids or gases into valuable products; for example, a mineral processing plant transforming crushed rock into a refined product. Power plants produce energy from fuels.

pumped hydro Energy storage arrangement in which water is pumped up to a high reservoir when excess electricity is available, and then released later through a turbine to generate electricity, to help meet peak power demand.

radioactive waste By-products of nuclear fission or other materials that release radiation and energetic particles, with gradually reducing levels of activity expressed in terms of half-life – the time needed for the level of activity to decrease by 50 per cent.

reliability Time performance of machinery or plant with no failures, usually measured as mean time between failures (MTBF).

renewable energy Energy derived from very large naturally occurring sources such as wind, solar radiation, tides, ocean waves and geothermal energy.

risk assessment Human decision-making process evaluating the likelihood and consequences of unpredictable but foreseeable events, to decide on control measures to reduce the likelihood and consequences of undesirable events.

unit operation (chemical engineering) Single step in a chemical engineering process; for example, separation of powder from liquid containing solid particles.

CHEMICAL ENGINEERING

30-second foundation

3-SECOND CORE
Chemical engineers design and operate process plants in many industries. They produce the materials needed by human societies, including many not considered to be 'chemicals'.

3-MINUTE IDEA
Chemical engineering is critical for health in providing supplies of safe drinking water, treating sewage and industrial effluent. Engineered water supplies and sanitation lead to greater health improvements than medical advances. Food packaging and processing greatly reduces waste, reducing the amount we need to grow. Many chemical plants are potentially very dangerous if operated incorrectly, therefore, chemical engineers devote a large part of their efforts to ensuring the safety of these large process plants.

Chemical engineers design and run the process plants making materials we need for human civilization. Water, fuel and medicines all come from such plants. Engineers understand how these complex arrangements of machines work together in a harmonized way. This understanding is a blend of practical knowledge, intuitive art, science and mathematics. The plant comprises an interlinked series of 'unit operations'. Engineers represent each unit with a mass and energy balance that specifies the energy and materials going in and out of a unit operation. Since matter and energy are neither created nor destroyed, chemical and thermodynamic equations determine where all the mass and energy goes. The mass balance guides creation of a process flow diagram (PFD), the starting point for design. The plant is defined by the piping and instrumentation diagram (P&ID), showing everything needed to control the process and make it safe to operate. A general arrangement (GA) drawing shows the actual plant layout, usually in three dimensions. Every detail is checked before construction. Process plants can cost hundreds of millions of pounds, and can be very dangerous if not designed and operated properly. Engineers need to ensure they have the right balance of cost, safety and reliability before construction begins.

RELATED TOPICS
See also
ORGANIZATIONAL SAFETY
page 82

PROCESS PLANT SAFETY
page 84

PLASTICS & FERTILIZERS
page 86

3-SECOND BIOGRAPHIES
GEORGE E. DAVIS
1850–1906
British engineer, now considered to be the father of chemical engineering, who invented key concepts such as unit operations.

MARGARET ROUSSEAU
1910–2000
American chemical engineer who designed the first commercial penicillin plant and was also the first female member of the American Institute of Chemical Engineers.

30-SECOND TEXT
Seán Moran

Engineers must test every part of the plant before full-scale production commences.

POWER GENERATION & ENERGY STORAGE

30-second foundation

Early engineers created city-wide power grids in the 1880s and every city relies on them today. Mechanical generators provide most electricity, driven by steam turbines heated by coal or nuclear reactors, or by gas turbines. The inertia of spinning turbines stabilizes the grid. Switching on a light creates an instantaneous increase in energy demand, slightly slowing a massive generator, and automatic systems then feed more gas or steam to compensate. Engineers have to make sure there is capacity to meet peak demand: many generators have to be kept spinning just in case they are needed. The second law of thermodynamics limits generation efficiency: generators produce carbon and other pollution and unwanted waste heat. Renewables – wind, solar and tidal – are helping to decarbonize electricity generation, although supplies are variable. Constantly adjusting power from mechanical generators to balance renewable supply with electricity demand is expensive and creates more pollution. 'Grid scale energy storage' – giant batteries – are becoming economically feasible. Water electrolyzers produce hydrogen to be stored and used later in fuel cells, which can compensate for renewable power and demand variations.

3-SECOND CORE

Renewable energy supply is intermittent. Energy storage from huge batteries smooths out supply and demand variations.

3-MINUTE IDEA

Lithium-ion batteries were originally developed for consumer electronics – mobile phones and laptops. Engineers are increasingly using them to power cars and store energy for the grid. In the struggle to keep electricity prices low, they place significant demands on large batteries, stimulating research into safer, cheaper batteries with improved storage. Made from hundreds of thousands of mass-produced thumb-sized cells, giant batteries offset their high cost by storing cheap surplus power and selling it back when demand and prices are high.

RELATED TOPICS

See also
WIND ENERGY
page 68

NUCLEAR POWER
page 80

ENERGY & FINANCE
page 140

3-SECOND BIOGRAPHIES

WILLIAM ROBERT GROVE
1811–96
Invented the gas battery in 1839; principles used by NASA.

NIKOLA TESLA
1856–1943
Developed alternating current generators, motors and electrical transmission systems.

EDITH CLARKE
1883–1959
Pioneered mathematical descriptions of power grids for predicting load-carrying capacity and stability.

30-SECOND TEXT

Paul Shearing

Electric energy keeps human societies running.

NUCLEAR POWER

30-second foundation

3-SECOND CORE
Nuclear power provides base load electricity and critical grid-stabilizing capacity. With minimal greenhouse gas emissions, many countries will rely on nuclear power to meet emission reduction targets.

3-MINUTE IDEA
Growing demands for environmental and biodiversity preservation, maintaining indigenous peoples' rights and reducing greenhouse emissions are all factors motivating engineers to develop small modular nuclear reactor power plants (SMRs). Reactor modules can be quickly exchanged before the fuel is exhausted and taken to a central refurbishing factory for refuelling and waste containment. SMRs will be safer and easier to manage, and will provide base load power while stabilizing power grids.

Nuclear fission in power reactors produces heat to generate electric power. Power reactors use very little fuel – mainly uranium – about 50 million times less volume than fossil fuel power stations for the same energy. Highly radioactive waste products are contained; the nuclear power industry is the only one that stores and processes all its waste. As with fossil-fuel power stations, nuclear power stations work best producing base load power, running continuously. By keeping enough generation capacity in reserve, system controllers can access extra capacity for demand variations. Even with ongoing safety concerns, nuclear power will be needed in many countries to help maintain power supplies and stabilize grid frequency with spinning inertia while reducing greenhouse gas emissions. Clean renewable power can gradually take over from coal and gas, but storing enough electricity in batteries or pumped hydro reservoirs to match supply variability is expensive and needs new technologies. If controlled nuclear fusion is successfully achieved, fusion reactors might one day produce vast energy supplies. However, the technology will still need many decades of further development.

RELATED TOPICS

See also
POWER GENERATION & ENERGY STORAGE
page 78

ORGANIZATIONAL SAFETY
page 82

ELECTRICAL ENGINEERING
page 94

3-SECOND BIOGRAPHIES

SIR CHRISTOPHER HINTON
1901–83
Led the design of Calder Hall, the first civilian nuclear power station.

ENRICO FERMI
1901–54
Built the first nuclear fission research reactor, in Chicago; fermium was named in his honour.

YANOSUKE HIRAI
1902–86
Oversaw design of nuclear power plants in Japan, which withstood the 2011 earthquake.

30-SECOND TEXT
Jorge Spitalnik

Nuclear power reactors produce negligible greenhouse gases.

100
$5f^{12}7s^2$
Fm
Fermium
257.095

ORGANIZATIONAL SAFETY

30-second foundation

Complex nuclear power plants and oil refineries are very reliable, but when they fail, the results can be disastrous for workers, the public, the environment and the corporation. Keeping them safe requires more than technical excellence. Disaster investigations rarely reveal new technical knowledge, but rather highlight how and why existing technical knowledge was not applied. Management systems are procedures and standards that reflect the best way of designing, operating and maintaining hazardous facilities. Formal risk assessment ensures that hazards are identified and risks are mitigated. A systems view of factors causing disasters requires that we look beyond the technology to the people and the organizational environment. Engineers avert disasters by reporting 'near misses' – minor events that might otherwise have led to catastrophes. However, as employees, their choices are influenced by organizational factors, for example who reports to whom and supervisors' key performance indicators. Engineers also have ethical and professional values that help inform their practice, especially when it comes to pushing bad news up through an organization that could help avert catastrophes.

3-SECOND CORE

Organizational safety requires engineers to be aware about the social factors that keep complex plants safe. High-reliability organizations adopt human behaviour strategies to avoid disasters.

3-MINUTE IDEA

Disasters are rare events, so the enemy of disaster prevention is complacency. It is helpful for engineers who work with complex hazardous systems to maintain an excellent safety imagination, an ability to anticipate a chain of events that could cause a disaster. Good engineers avoid psychological rigidities that make it difficult to see how everyday actions can contribute to disaster. They listen to plant operators and maintainers who can notice early warning signs of impending faults.

RELATED TOPICS

See also
NUCLEAR POWER
page 80

PROCESS PLANT SAFETY
page 84

FLOATING FACTORIES
page 120

3-SECOND BIOGRAPHIES

KARL WEICK
1936–
Influenced organizational safety with work on high-reliability organizations.

JAMES REASON
1938–
Known for the 'Swiss Cheese' model (how accidents can happen even with robust safety measures in place).

JUDITH HACKETT
1954–
Led developments in health, safety and environmental regulations for chemical industries and process plants.

30-SECOND TEXT

Jan Hayes

Many high-reliability organizations provide helpful models for effective safety.

PROCESS PLANT SAFETY

30-second foundation

Chemical process plants have to make money for their owners. However, safety, health and environmental (SHE) issues are even more important for chemical engineers. Chemical process plant accidents have the potential to cause tens of thousands of deaths, which is why engineering codes of ethics, such as those of the American Institute of Chemical Engineers (AIChE), demand that engineers consider safety before profits. Regulatory bodies and governments can order plants to be shut down if they are not operated safely, putting profits at risk. A large accident can destroy a plant, eliminating income and profits. Chemical engineers do not aim for perfect safety: they know that every improvement adds cost. Regulations use terms such as 'as low as reasonably practicable' (ALARP) and 'so far as is reasonably practicable' (SFAIRP) to define the required standards of safety. Engineers are not required to make plants safer if the cost of safety improvement is grossly disproportionate to the benefit gained. Tools such as hazard Identification (HAZID), hazard analysis (HAZAN) and hazard and operability study (HAZOP) help to detail every safety risk and design intrinsically safe process plants.

3-SECOND CORE

Chemical plants can contain large amounts of dangerous materials. Great care is taken by plant designers and operators to make sure these do not cause harm.

3-MINUTE IDEA

It is important to distinguish between process safety, which is about managing the safety of large quantities of hazardous materials on a process plant, and personnel safety, which is about protecting individual workers. Process safety controls the risk of fatalities associated with large quantities of flammable, explosive or toxic materials. If uncontrolled, thousands may become seriously ill or die, as happened at Bhopal in 1984 and Seveso in 1976.

RELATED TOPICS

See also
CHEMICAL ENGINEERING
page 76

ORGANIZATIONAL SAFETY
page 82

FLOATING FACTORIES
page 120

3-SECOND BIOGRAPHIES

ALICE HAMILTON
1869–1970
American physician and first female faculty member at Harvard University who became the leading US authority on lead poisoning, industrial toxicology and occupational health and safety.

TREVOR KLETZ
1922–2013
British chemical engineer credited with introducing the concept of inherent safety, and a major promoter of HAZOP methods for safe design.

30-SECOND TEXT

Seán Moran

Engineers anticipate mistakes by operators and maintainers so they can design plants that are safe.

PLASTICS & FERTILIZERS

30-second foundation

3-SECOND CORE

Cheap synthetic plastics and fertilizers help sustain modern society but have caused unacceptable pollution. With appropriate incentives, chemical engineering companies can find alternatives to reduce pollution.

3-MINUTE IDEA

Thermoplastic polymers are light, strong, cheap and easily moulded into intricate shapes. Mineral fertilizers remain essential for large-scale food production. Alternatives are more expensive, but do reduce waste and/or pollution. Political action could impose 'external costs' to change the economics of plastics and fertilizers, encouraging chemical engineers to develop lower cost alternatives. Governments can encourage less polluting alternative plastics and fertilizers.

The production of nitrogen fertilizers using nitrogen from the air and hydrogen from oil or gas is one of chemical engineering's great successes. In 1909, German chemist Haber produced ammonia (the raw material of nitrogen fertilizers) in the lab. Collaboration with Bosch created an industrial process, which today yields 450 million tonnes of ammonia-based fertilizer annually. Ammonia fertilizers and insecticides have quadrupled the productivity of farm land. Chemist Leo Baekeland produced the first engineered plastic, Bakelite, in 1907, a liquid that sets in a mould when heated. This robust insulator transformed electrical engineering. Most plastics today are thermoplastic polymers: long chains of small molecules produced from oil and natural gas in petrochemical plants, heated for moulding or extruded as thin sheets. Environmental pollution from fertilizers and plastics is controversial. Nitrogen fertilizers dissolve in water and drain into rivers; synthetic plastics break down slowly, contaminating the oceans. Recycled plastic is more expensive to produce and of a lower quality than new material. Biofertilizers made from waste are more expensive than industrial chemical fertilizers. Regulation and taxes are needed, therefore, to encourage plastics recycling and biofertilizer production.

RELATED TOPICS

See also
CHEMICAL ENGINEERING
page 76

RESOURCE SCARCITY
page 142

FEEDING OUR WORLD
page 144

3-SECOND BIOGRAPHIES

DIANNE DORLAND
1948–
American chemical engineer and first female president of the American Institute of Chemical Engineers, honoured for her work on reducing mercury pollution in paper manufacturing and on engineering education.

ISATOU CEESAY
1972–
Known as the 'Queen of Recycling' in The Gambia, Ceesay developed women's community plastic recycling schemes.

30-SECOND TEXT

Seán Moran

Despite being a success story, fertilizers and plastics have proven to be hazardous to the environment.

N
H
H
H

9 December 1868
Fritz Haber born in Breslau, Prussia

27 August 1874
Carl Bosch born in Cologne, Germany

1891
Haber receives doctorate after studying at the Technical College of Charlottenburg

1892
Haber joins University of Jena after further studies in Zürich

1894
Haber joins University of Karlsruhe, researches dye technology and electrochemistry, and catalytic formation of ammonia

1898
Bosch receives doctorate in organic chemistry from University of Leipzig

1899
Bosch joins BASF as chemistry technologist

1905
Haber publishes influential book on chemical thermodynamics of gas reactions

1906
Haber appointed professor

1908
BASF hires Haber to lead design of a commercial high-pressure ammonia synthesis process

1909
Bosch evaluates commercial potential

1914
Bosch supervises construction of first large-scale factory

1915
Haber joins army and leaves BASF to work on poison gas production

1916
BASF boosts explosive production from 36,000 tonnes a year to 160,000 tonnes by 1918

1925
Bosch appointed to lead IG Farben

1931
Haber & Bosch awarded the Nobel Prize with Friedrich Bergius

1933
Haber leaves Germany for Cambridge

29 January 1934
Haber dies in Basel

1935
Bosch appointed IG Farben board chairman

26 April 1940
Bosch dies in Heidelberg

FRITZ HABER & CARL BOSCH

A century ago, Fritz Haber and Carl Bosch created the industrial process that provides the nitrogen fertilizers sustaining most of the world's food production today. Theirs is a story of triumph and tragedy.

Both men were born into industrialist families (Bosch as the nephew of Robert Bosch, founder of Robert Bosch GmbH) at a time when chemistry was the advanced technology of the age, and both studied at technical universities. By 1900, chemicals such as dyes, nitrates and ammonia were crucial for industries and defence. Britain and France relied on colonies for supplies; Germany turned to its chemists.

Haber, with his industrial chemistry training, had realized the potential for synthesizing ammonia. Working with his assistant, Robert Le Rossignol, he found that heat and pressure could induce nitrogen and hydrogen to form ammonia in the presence of a catalyst. After 14 years of research, Haber persuaded the chemical company BASF to take an interest. They appointed Bosch, by then an experienced chemical engineer, to evaluate the commercial potential.

Bosch led a team that performed over 20,000 tests in a few months to find the best catalyst, before starting work on a full-scale production plant. The outbreak of war in 1914 was a temporary setback. Haber secured military support by promoting the possibility of making low-cost explosives. Whilst Bosch worked to scale up production, Haber joined the army to produce and deploy poison gas. Tragically, his wife committed suicide days later. Bosch and his engineers constructed two large factories to produce ammonia and nitrate explosives. In 1918, production had increased sufficiently to allow fertilizer manufacture, credited with averting famine. After the war, Bosch led BASF's commercial developments and co-founded IG Farben, dominating the world's chemical industry.

Haber was born Jewish but converted to Christianity, seeking approval in German society. Despite his support for the German war effort, he was forced to resign his university post as Hitler became Chancellor, and Cambridge scientists helped him leave Germany. He died on his way to start a new institute in Palestine. Bosch was outspokenly critical of Nazi policies and was gradually relieved of his position, became dependent on alcohol and died a few years later.

Some see Haber and Bosch as creators of the German military-industrial complex, providing the technology that enabled Germany to continue World War I by several years. Others hail the ammonia fertilizer production process a triumph that sustains food production for the world's population today – indeed, the Haber–Bosch process produces 450 million tonnes of nitrogen fertilizer every year and has vastly increased agricultural yields on a global scale.

James Trevelyan

ELECTRICAL & ELECTRONIC ENGINEERING

ELECTRICAL & ELECTRONIC ENGINEERING
GLOSSARY

avionics Electronics specially designed for use in aircraft and spacecraft.

base load power Amount of electricity generation that is required continuously, 24 hours a day, to meet a variable demand. Additional 'peak load' power is needed as demand rises from the 'base load' level.

binary Number represented by a string of 1s and 0s instead of normal decimal digits; used by digital computer designers.

chip (electronic engineering) Integrated circuit.

cochlear implant Electronic device that stimulates nerves which would respond to sound in a healthy human ear, partially restoring hearing for some deafness.

complex number Numbers with two components – one real and one imaginary – that enable compact representation of periodic signals, making analysis of electric circuits much easier.

debug Find design and coding mistakes in computer programs.

demand (electrical engineering) Requirement for electricity to be provided, determined by the number and power requirements of all devices currently connected and switched on in an electricity supply network.

electromagnetic radiation Radiation associated with changing electric currents – depending on the frequency, this radiation appears as radio waves, microwaves, infrared, visible light, ultraviolet, x-rays or gamma rays, all travelling at the speed of light.

filter (electronic engineering) Electronic circuit or digital computer program routine used to modify the characteristics of a signal.

fluorescent Material that glows when exposed to invisible ultraviolet radiation.

frequency hopping Radio transmitter and receiver with synchronized and rapid frequency changes, making it very difficult to intercept the transmission. Typically used by security forces.

fuel cell Reaction vessel or cell in which gases and liquids combine and generate electricity.

grid (electrical engineering) Entire electric power supply system, including electricity-generating stations, transformers, switch yards and powerline interconnections.

integrated circuit Single piece of semiconductor material on which many circuit elements have been fabricated, with connections and encased in a plastic or ceramic insulating package.

lithography Process used to fabricate transistors and other circuit elements on the surface of a semiconductor.

microgrid Small network of connected electricity users with local electricity generation, often with renewables, that can operate independently or draw power from the main grid when needed.

Moore's Law Named after Gordon Moore, co-founder of Intel, who predicted the number of transistors on an integrated circuit chip would double every two years, later amended to 18 months.

optical fibre Extremely clear glass fibre used to transmit digital signals over long distances with pulses of light.

pacemaker Electronic device that stimulates nerves controlling heart muscles; used to stimulate regular heartbeats in patients with irregular or intermittent heartbeats.

pumped hydro Energy storage arrangement in which water is pumped up to a high reservoir when excess electricity is available, and then released later through a turbine to generate electricity, helping to meet peak power demand.

radar Transmission of high-frequency radio signals and detection of echoes: the time taken for the echo to come back indicates the distance of the reflecting object.

renewable energy Energy derived from naturally occurring sources such as wind, solar radiation, tides, ocean waves, biomass and geothermal energy.

shielding Metal enclosure to reduce unwanted effects of radiation.

sonar Transmission of high-frequency sound and detection of echoes: the time taken for the echo to come back indicates the distance of the reflecting object.

transistor Active semiconductor circuit element enabling a small current to influence a much larger current in another circuit branch. Used as a switch or amplifier, depending on the application.

ELECTRICAL ENGINEERING

30-second foundation

Abstract mathematical models and physics frame engineers' ideas about electricity: invisible flow of electric charge through conductors and associated electromagnetic fields. These concepts enable electrical engineers to create electric energy generation and transmission systems that provide us with light, heat and power when we need it. They also enable electronic engineers to create wireless communications, computers and mobile phones. Practical means for storing electric energy require transformation to and from chemical energy in batteries, fuel cells, mechanical energy in rotating machines and pumped hydro storage reservoirs. All are expensive and some energy is lost, so the primary concern for electrical engineers is regulating electricity generation and transmission to meet the demand. Every time an energy-saving light or even a huge electric arc furnace is switched on or off, a generator has to provide more or less electricity instantly, and then countless other adjustments, mostly automatic, compensate for the resulting change in energy flow. Electricity is dangerous for people and animals, so electrical engineers are also preoccupied with managing numerous safety hazards and automatic protection devices that switch off power instantly when needed.

3-SECOND CORE
Electrical engineers sustain electric energy generation and distribution, providing critical infrastructure for human civilization. They have to match supply with demand and ensure safety for humans and animals.

3-MINUTE IDEA
Centralized energy grids rely on strong societal governance, trust and cooperative behaviour. The tension between supply reliability, the willingness of users to pay, safety, and protecting our environment has never been easy to manage. Engineers are exploring alternatives such as intelligent community-scale microgrids, partly to manage fluctuating renewable energy supplies. Increasingly, managing users' behaviour is critical: the notion of an infinite supply of electricity whenever it is needed does not match reality.

RELATED TOPICS
See also
WIND ENERGY
page 68

POWER GENERATION & ENERGY STORAGE
page 78

3-SECOND BIOGRAPHIES

CHARLES-AUGUSTIN COULOMB
1736–1806
Investigated electrostatic charge and forces between nearby objects, and similar phenomena in magnetism.

HANS CHRISTIAN ØRSTED
1777–1851
Discovered the relationship between electric currents and associated magnetic fields.

GEORG SIMON OHM
1789–1854
Discovered that the voltage difference along a conductor is proportional to the electric current.

30-SECOND TEXT
James Trevelyan

Power grids support civilization; engineers make sure that they are safe, economic and dependable.

ELECTRONIC ENGINEERING

30-second foundation

3-SECOND CORE
Electronic engineers create circuits in which tiny electrical components convey, transform and amplify sounds, images and other kinds of information, enabling many other technologies.

3-MINUTE IDEA
In the early twentieth century, engineers discovered that complex numbers, devised by mathematician Gerolamo Cardano centuries earlier, could be used to represent electrical signals. This discovery is the basis of the stability of circuits and electric power grids, and of broadband technologies. Complex numbers provide the foundation for circuit theory, enabling modern radar, ultrasound and the analysis of filter circuits. They unify the ideas used by all electrical and electronic engineers.

The 1903 invention of vacuum tubes, also known as valves, saw the dawn of electronic engineering. Valves gave way to solid-state transistors in the 1960s. Soon, integrated circuit 'chips' (ICs) enabled faster, more accurate and, eventually, far more complex circuits. Valves, transistors and ICs are all 'active devices'. Voltage applied to passive devices such as resistors or lamps only influences current through the device. However, voltage applied to an active device influences current in a different circuit branch. This 'trans-action' was used to coin the name transistor, from 'trans-resistor'. Electronic engineers created broadcast radio, which in turn led to the greatest social upheaval in the twentieth century. Electronic engineers helped automate factories, created radar, sonar, computers, pacemakers and cochlear implants, and even electronic musical instruments. They perfected sound reproduction, television, global positioning systems, mobile phones and the Internet. Electronics turned wind-up wristwatches and film cameras into antique curiosities. Electronic engineering advances leading to electric vehicles, faster, cheaper computers and ubiquitous communications are shaping life in the twenty-first century, just as radio communication and broadcast radio changed twentieth-century societies.

RELATED TOPICS
See also
MECHATRONICS
page 62

COMPUTER ENGINEERING
page 98

SIGNAL PROCESSING
page 108

3-SECOND BIOGRAPHIES

JAMES CLARK MAXWELL
1831–1879
Built a mathematical theory for electromagnetic radiation, providing the concepts behind wireless communication.

WILLIAM SHOCKLEY
1910–89
Led the development of the transistor at Bell Labs and was awarded a Nobel Prize.

HEDY LAMARR
1914–2000
Helped invent a frequency-hopping military radio to defeat interception and jamming. The same technology is used in mobile broadband.

30-SECOND TEXT
Jonathan Scott

Electronic devices and computers enable many other new technologies.

FIRE
2
3
4
5
4
3
2

COMPUTER ENGINEERING

30-second foundation

Inside a computer or phone you can find small, black 'chips' up to 2 cm (¾ in) in size, mounted on circuit boards. Inside each chip is a small, densely packed electrical circuit, known as an 'integrated circuit' (IC). The first transistor radios had circuit boards with individual transistors in tiny metal cans with three metal legs. Engineers realized that making hundreds of transistors on a single piece of silicon was possible, and soon the first integrated circuits appeared, with up to ten transistors each. Transistors work well as on/off switches, and engineers realized that they could use transistors to construct complex logic circuits to add and subtract numbers represented as strings of 1s and 0s – binary. Arrays of transistors on ICs could perform lightning-fast binary calculations. Intel, one of Silicon Valley's first IC manufacturers, even derived its name from the term 'integrated electronics'. Computer engineers design and fabricate even denser and faster computer chips with smaller and smaller elements. Reliably replicating features just a few atoms in size is one challenge; another is electromagnetic interference; as circuit elements are more closely packed, radio signals in one element can interfere more easily with an adjacent element, so additional shielding features are needed.

3-SECOND CORE
Computer engineering enables the design and fabrication of enormously complex electronic circuits to process information represented as binary numbers.

3-MINUTE IDEA
Smaller chips mean faster and cheaper circuits. Shrinking transistors and cramming more transistors together has limits because silicon atoms are finite in size. Engineers are changing how transistors are manufactured, by using precise optical lithography techniques. Circuits are patterned using extreme ultraviolet light, allowing for seven nanometre feature sizes. Other semiconductor materials, such as silicon-germanium, are used to make transistors. Innovative manufacturing has defied the physical limits of the transistor's atoms.

RELATED TOPICS
See also
ELECTRONIC ENGINEERING
page 96

SOFTWARE ENGINEERING
page 100

INFORMATION & TELECOMS
page 106

3-SECOND BIOGRAPHIES

JOHN VON NEUMANN
1903–57
Helped establish practical designs for digital computers and developed programs.

KONRAD ZUSE
1910–95
Created the world's first programmable digital computer in 1941.

ALAN TURING
1912–54
Developed the logical foundations for programmable computers and what would later be known as 'software'.

30-SECOND TEXT
Kate Disney

Computers with billons of transistors are now routinely fabricated on single chips.

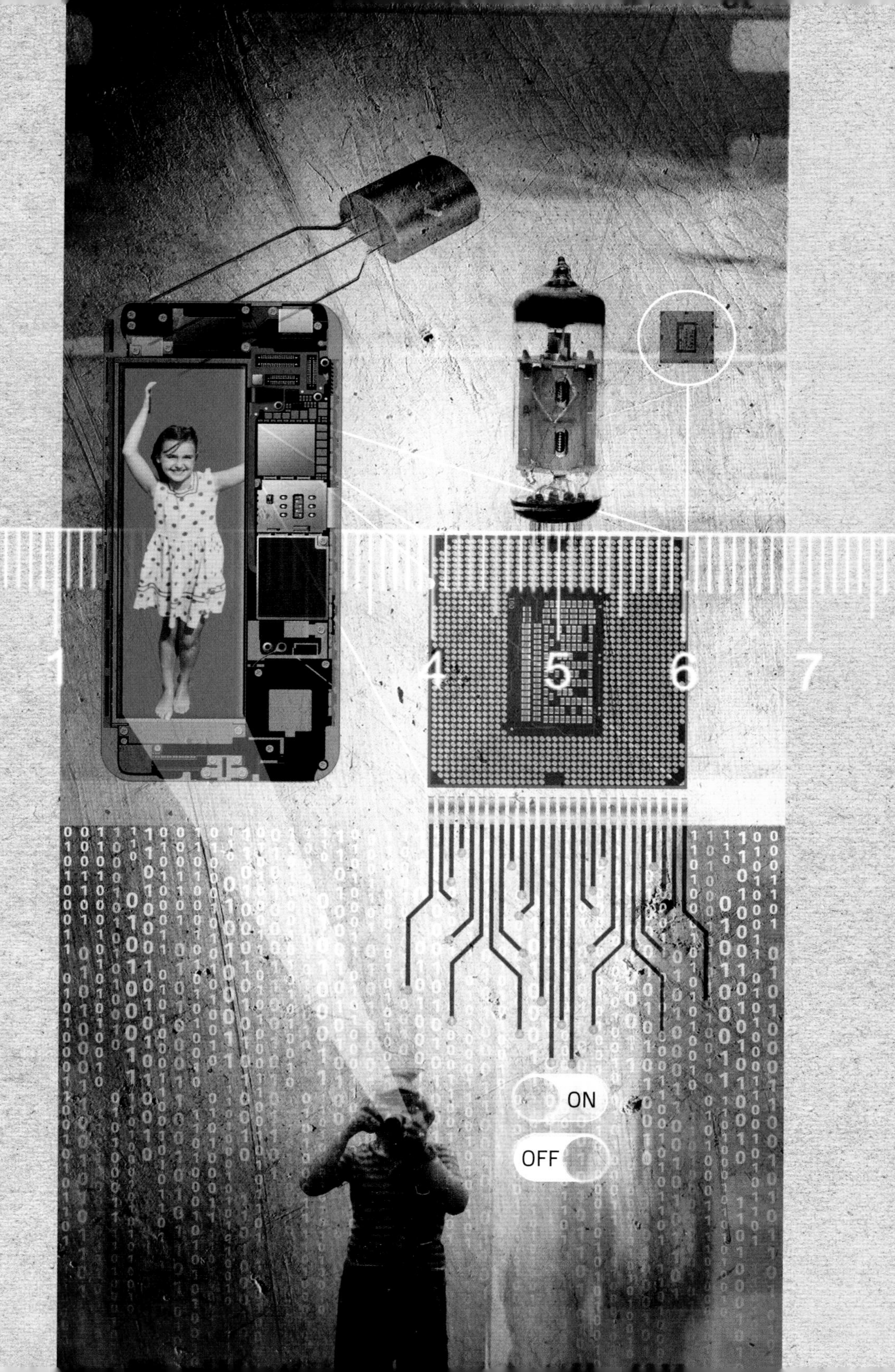
0 1 4 5 6 7 8
m
ON
OFF

SOFTWARE ENGINEERING

30-second foundation

Software engineers create, maintain and develop software systems: the instructions and data that enable computers to perform useful tasks. Software systems are diverse, from video games all the way through to surgical robots. Engineers follow the same processes every time, but the emphasis varies according to the importance of ensuring that there are no critical defects. They start with the requirements – what the software must do and how it interacts with people, machines and other systems. They devise tests to verify that the software performs as expected. They construct computer models representing the software so they can predict performance, or prove its logical correctness. Next, engineers interpret requirements to write programs in computer languages. They encode algorithms into the software: known methods for achieving common tasks, such as sorting a list of names into alphabetical order. Computer languages allow programmers to write human-readable instructions which a computer translates for the processor that will run the software. Most software includes a user interface (UI), enabling someone to interact with the system. Engineers rely on software tools, programs that write much of the software automatically. Finally, they perform tests to eliminate mistakes (bugs).

3-SECOND CORE
Software engineers create the programs that enable computers to perform useful tasks. Much of the effort goes into creating tests to find programming errors.

3-MINUTE IDEA
Software architecture refers to system design. As with a physical building, there are many different architectural styles with different characteristics, ranging from a single program to multiple separate programs called components or services. Choosing appropriate architecture for particular requirements demands judgement and experience; once programming has started, it can be hard to change. The right architecture can make it easier to allow changes needed late in a project.

RELATED TOPICS
See also
COMPUTER ENGINEERING
page 98

INFORMATION & TELECOMS
page 106

DRIVERLESS CARS
page 128

3-SECOND BIOGRAPHIES

ADA LOVELACE
1815–52
English mathematician who recognized that Babbage's mechanical computer, or 'analytical engine', had applications beyond pure calculation, and published the first algorithms.

EDSGER WYBE DIJKSTRA
1930–2002
Dutch computer scientist who formalized many critical ideas in computer science, such as compilers, and helped invent structured programming languages such as Pascal.

30-SECOND TEXT
Andrew McVeigh

Devising tests to detect defects (bugs) takes time, but is critical for good-quality software.

9 December 1906
Born in New York, USA

1928
Graduates in physics and mathematics from Vassar College

1931
Starts teaching mathematics and science at Vassar College

1934
Graduates with a PhD in mathematics from Yale, continues teaching at Vassar College

1943
Enlists with US Navy Reserve

1944
Assigned to Mark 1 computer team at Harvard University

1949
Joins Eckert–Mauchly Computer Corporation

1952
Leads development of A-0, a 'linking loader', which she called a compiler, to automatically combine standard program components (subroutines) into a working program

1954
Becomes director for programming development

1959–66
Advises panels of experts on development of COBOL

1966
Retires from US Navy Reserve

1967
Recalled by the Navy, leading a team developing validating programs

1973
Promoted to Captain

1983
Promoted to Commodore

1985
Promoted to Rear Admiral

1986
Retires from Navy, becomes senior consultant at Digital Equipment Corporation

1990
Retires from DEC

1 January 1992
Dies, buried in Arlington National Cemetery

GRACE BREWSTER HOPPER

The safety of money in a bank account reflects the work of engineers and the leadership of Grace Hopper, who insisted that computers could be programmed using text resembling human languages. She influenced the development of COBOL, a programming language used from 1960 to today.

The young Hopper satisfied her curiosity for machines by dismantling alarm clocks. Inspired by her mother's passion for mathematics, she studied the subject along with physics at Vassar College – founded to promote the education of women – and then completed a PhD in mathematics at Yale in 1934, whilst teaching at Vassar College.

After the 1941 attack on Pearl Harbor, Hopper was determined to follow in the footsteps of her great grandfather, an admiral in the US Navy. She enlisted in the Navy reserve and graduated from Northampton Midshipmen's School as lieutenant. She was told she was underweight, too old (38) and, being a professor, too valuable for combat duty. Instead, she was assigned to the Harvard University team working on electro-mechanical computers using relays and electro-mechanical counters – common components of telephone exchanges until the 1970s.

In 1946 Hopper applied to join the Navy but was refused again. She joined Eckert–Mauchly, a private company formed from the team that developed the world's first electronic computer, ENIAC. She remained with the Navy reserve.

Hopper recognized that programming computers was difficult to learn because knowledge of electronic circuits and mathematics was needed to translate requirements into a sequence of instructions written in 1s and 0s. She demonstrated that many more people could learn to program computers if the instructions consisted of human language words such as ADD, SUBTRACT, REPEAT and INDEX. Standardizing newly emerging computer languages was also critical, requiring negotiation skills to balance competing human and commercial interests.

Between 1959 and 1966, Hopper advised experts from US industry and government on development of a standard business-oriented data-processing language, COBOL, used for an estimated 80 per cent of all software by the 1990s. In 1966, Hopper was informed by the Navy for the third time that she was too old, and so retired – only to be recalled a year later to lead a team developing validating programs, ensuring COBOL provides the same results on different computers and operating systems. A six-month temporary term lasted for 20 years.

Hopper's ideas were adopted for every programming language that followed. She became a passionate advocate for using computers and also for the US Navy, giving hundreds of lectures every year until her death in 1992.

James Trevelyan

NANOTECHNOLOGY

30-second foundation

3-SECOND CORE
Nanotechnology enables engineers to create extremely small structures, some made from just a few atoms. Improved medical diagnosis methods and new materials are emerging as a result.

3-MINUTE IDEA
It can be hard for nanotechnology engineers to see their creations. Electron microscopes can distinguish features as small as 0.1 nanometre. To see individual atoms, engineers have created atomic force microscopes. A fine needle probe with just a single atom at the tip moves over a sample and sensors record the forces felt from features as small as individual atoms in the sample. With such high magnification, it can be difficult to work out where to start looking.

Nanotechnology engineers work with very small structures ranging from nanometres to microns in size. A human hair is about 90 microns in diameter (90,000 nanometres). At this scale, electrostatic forces are far larger than the forces that we feel, like gravity. One application is cancer diagnosis, which today relies on surgical biopsy and expensive imaging scanners. Some cancer cells break away from tumours and circulate in the blood, developing a negative electrostatic charge on their surface. Engineers created positively charged fluorescent nanoparticles from magnetic iron oxide about the same size as large molecules (50–100 nanometres) but much smaller than cancer cells (20–30 microns). These particles attract themselves electrostatically to tumour cells in a blood sample. A magnet can then separate the tumour cells, allowing them to be identified as the nanoparticles glow under ultraviolet light. Engineers are also making new materials with special properties. For example, thin layers of nanoparticles can protect people from harmful sunburn and skin cancer. By using a combination of integrated circuit fabrication methods with electrochemical etching, engineers can mass-produce tiny sensors such as accelerometers and inertial sensors for the automobile industry at low cost.

RELATED TOPICS
See also
ELECTRONIC ENGINEERING
page 96

COMPUTER ENGINEERING
page 98

BIOMEDICAL ENGINEERING
page 110

3-SECOND BIOGRAPHIES

NIELS BOHR
1885–1962
Provided a fundamental understanding of the structure of atoms and molecules.

MAX KNOLL
1897–1969
Showed that a scanning electron beam could produce the image of an object in a vacuum.

MANFRED VON ARDENNE
1907–97
Demonstrated the first high-magnification electron microscope, amongst many other inventions.

30-SECOND TEXT
Donglu Shi

Nanoparticles can be used to detect tumour cells circulating in the blood stream.

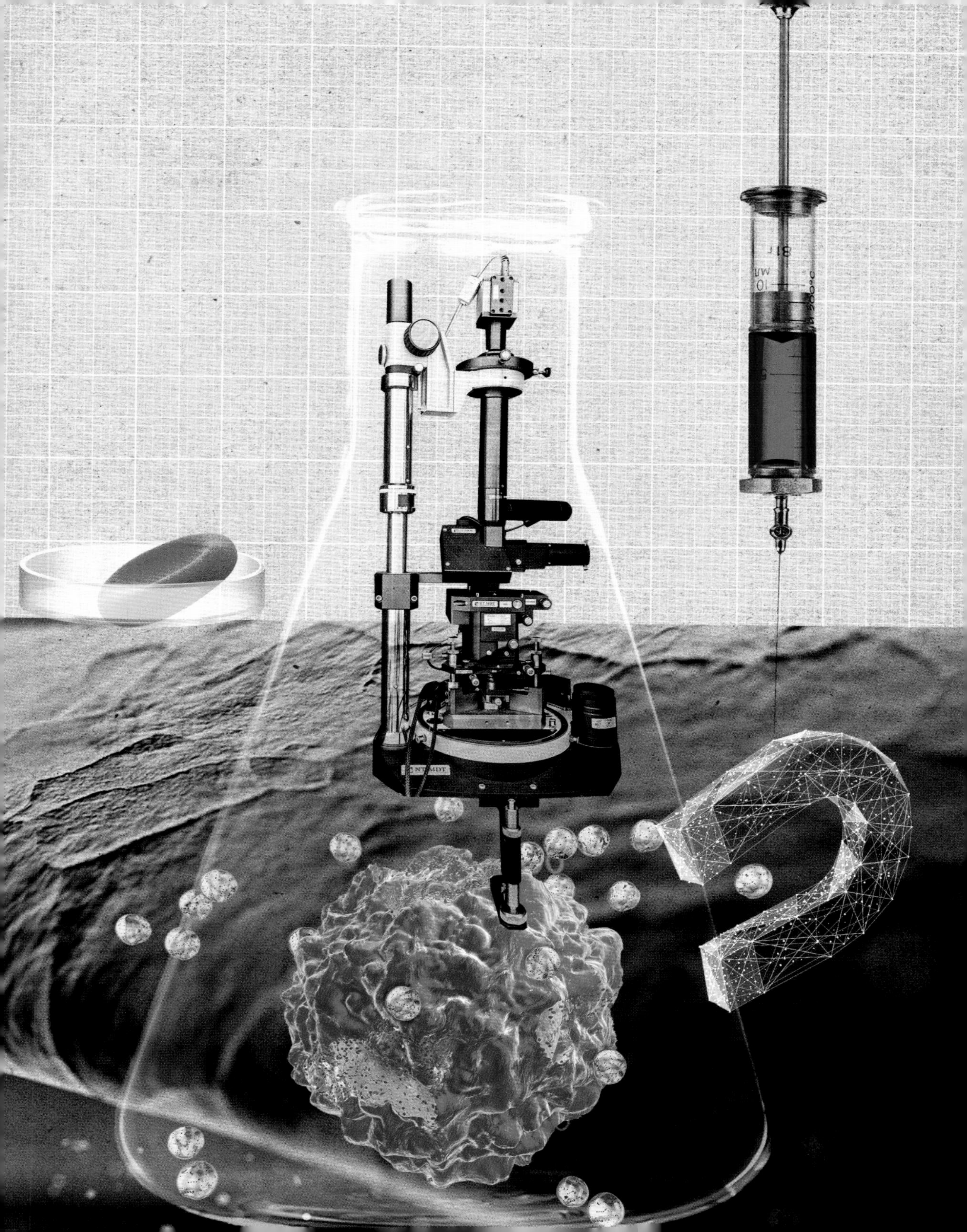
NT-MDT

INFORMATION & TELECOMS

30-second foundation

Information and Communications Technology (ICT) enables the rapid processing and communication of information, and reliable storage. Over the past few decades, engineers have created information networks connecting terminals such as computers, mobile phones, satellites, sensors and controllers by optical fibres, undersea cables and radio communication links. ICT engineers make sure that information flows through these networks with the required quality, at a reasonable cost. Information from the natural and societal world, originally in the form of data, text, sounds and images, is converted into digital form so that it can be stored and treated by digital computers. Engineers use the Nyquist-Shannon sampling principle to ensure that no information is lost in this conversion process. The global information network is the Internet, which links billions of users all over the world, while the Transmission Control Protocol and Internet Protocol (TCP/IP) governs the way that computers send information from one to another and acknowledge that they have received it. Furthermore, digital signal processing (DSP) techniques enable engineers to enhance, protect, encrypt or compress information – for example, filtering sounds to remove unwanted background noises.

3-SECOND CORE

ICT engineers create systems that enable information to be stored and transmitted anywhere, almost instantly, using special techniques to minimize the cost.

3-MINUTE IDEA

As a platform with various services ranging from email to e-library, e-learning, e-community, e-commerce, e-home, e-hospital, e-bank, e-manufacturing and so on, the Internet has gathered and stored a huge amount of data, which increases daily. 'Big data' refers to technology for searching, indexing, analysing and classifying data so that people can make use of it. Data collection raises important ethical questions for engineers because people are not always aware of the information about them that is being stored.

RELATED TOPICS

See also
COMPUTER ENGINEERING
page 98

SOFTWARE ENGINEERING
page 100

SIGNAL PROCESSING
page 108

3-SECOND BIOGRAPHIES

ALEXANDER GRAHAM BELL
1847–1922
Inventor of the telephone and founder of the huge Bell telephone company, now AT&T.

GUGLIELMO MARCONI
1874–1937
Worked on the long-distance radio transmission over the Atlantic Ocean in 1896, which saw the dawn of wireless communication.

CHEN ZUONING
1957–
Led the development of Chinese super-computers, the fastest in the world.

30-SECOND TEXT

Gong Ke

ICT has become such a powerful influence that many call this period the 'Information Age'.

SIGNAL PROCESSING

30-second foundation

Data from cameras and laser radar enables a driverless car to calculate its location and monitor nearby cars, pedestrians, bicycles and other potential obstacles. Twenty-five times a second, computers analyse gigabytes of images and radar reflections, extracting maybe 1,000 bytes of useful information. Tiny computers inside noise-cancelling headphones suppress ambient sounds so that people can enjoy music in noisy aircraft cabins, factories and offices. Hearing aids enhance sounds using similar methods. Cameras use signal processing to transform enormous amounts of data from image sensors into compressed images and video, providing economic ways to distribute films and entertainment using the Internet. Televisions use signal processing to improve the compressed images, getting the best possible results from low-priced displays. Engineers build mathematical models representing the behaviour of the interesting signals that might be buried in gigabytes of data, containing lots of irrelevant noise and other information. Then they construct filter algorithms that separate the interesting signals from all the other irrelevant data, often using software that designs the algorithms automatically.

3-SECOND CORE

Engineers use signal processing methods to separate useful information from irrelevant information and random noise, enabling economic transmission of films, images and sounds.

3-MINUTE IDEA

In a stunning series of insights, Kotelnikov, Whittaker, Nyquist and Shannon observed that a continuously variable signal could be sampled at twice the frequency of the fastest varying component, and fully reconstructed from just the sampled values. As human ears can only perceive sounds up to 20,000 Hertz, sampling at 40,000 Hertz is sufficient to provide a full-fidelity audio signal. This sampling theorem laid the foundations for digital signal processing technology, enabling worldwide telecoms and entertainment.

RELATED TOPICS

See also
ELECTRONIC ENGINEERING
page 96

COMPUTER ENGINEERING
page 98

INFORMATION & TELECOMS
page 106

3-SECOND BIOGRAPHIES

HARRY THEODOR NYQUIST
1889–1976
Identified the stability criterion for all feedback control systems and laid the foundations for Shannon's information theory.

VLADIMIR KOTELNIKOV
1908–2005
Discovered the Nyquist-Shannon sampling theorem.

CLAUDE ELWOOD SHANNON
1916–2001
Contributed to information theory and digital circuit design theory, cryptanalysis.

30-SECOND TEXT

James Trevelyan

Most electronic sounds, videos and images rely on signal processing in tiny computers.

BIOMEDICAL ENGINEERING

30-second foundation

The interdisciplinary biomedical engineering of the 1950s brought engineering solutions into medicine and biology, and is now one of the fastest-growing areas of technology. Physics inventions revolutionized medicine, such as von Helmholtz's ophthalmoscope in 1851, Rontgen's X-ray images in 1895, Einthoven's electrocardiogram in 1903 and Ruska and Moll's electron microscope in 1931. Engineering efforts led to the safe, affordable and reliable instrumentation we take for granted in hospitals, requiring knowledge of biology and medicine to complement electronic, optical and mechanical engineering. New, lightweight, highly corrosion-resistant materials, such as titanium, enabled engineers to design more elaborate, durable and comfortable prosthetic devices. In the 1970s, engineers extended X-ray imaging with computed tomography, and developed ultrasound imaging and nuclear magnetic resonance scanners. Electronics miniaturization enabled cochlear implants, at least partially restoring lost hearing. New materials, nanotechnology and 3D printing are now enabling novel methods to repair body tissues such as cartilage, bone, liver, kidneys, skeletal muscles, blood vessels and even the nervous system.

3-SECOND CORE

One of the fastest-growing fields of technology, biomedical engineering holds a prominent place as a means of improving medical diagnosis and treatment by healthcare professionals.

3-MINUTE IDEA

Biomedical engineers and regulators have developed stringent requirements for certification of medical devices, instrumentation and implants. Standards for hospital instrumentation, recently extended for home use, help ensure that instruments provide accurate information and operate safely. European directives such as RoHS2 govern the use of hazardous materials like radioactive isotopes. The high cost of medical devices reflects the work that engineers have to perform to demonstrate that they are safe to use.

RELATED TOPICS

See also
MATERIALS ENGINEERING
page 60

ELECTRONIC ENGINEERING
page 96

NANOTECHNOLOGY
page 104

3-SECOND BIOGRAPHIES

WILHELM RÖNTGEN
1845–1923
German mechanical engineer and physicist who discovered X-rays and invented the X-ray photograph, enabling doctors to see inside living human bodies for the first time.

WILLEM EINTHOVEN
1889–1976
Indonesian-born Dutch physiologist who invented the first ECG, measuring extremely small electrical signals from a beating heart.

30-SECOND TEXT

Hung Nguyen

Biomedical engineers have created new instruments, implants and methods for tissue repair.

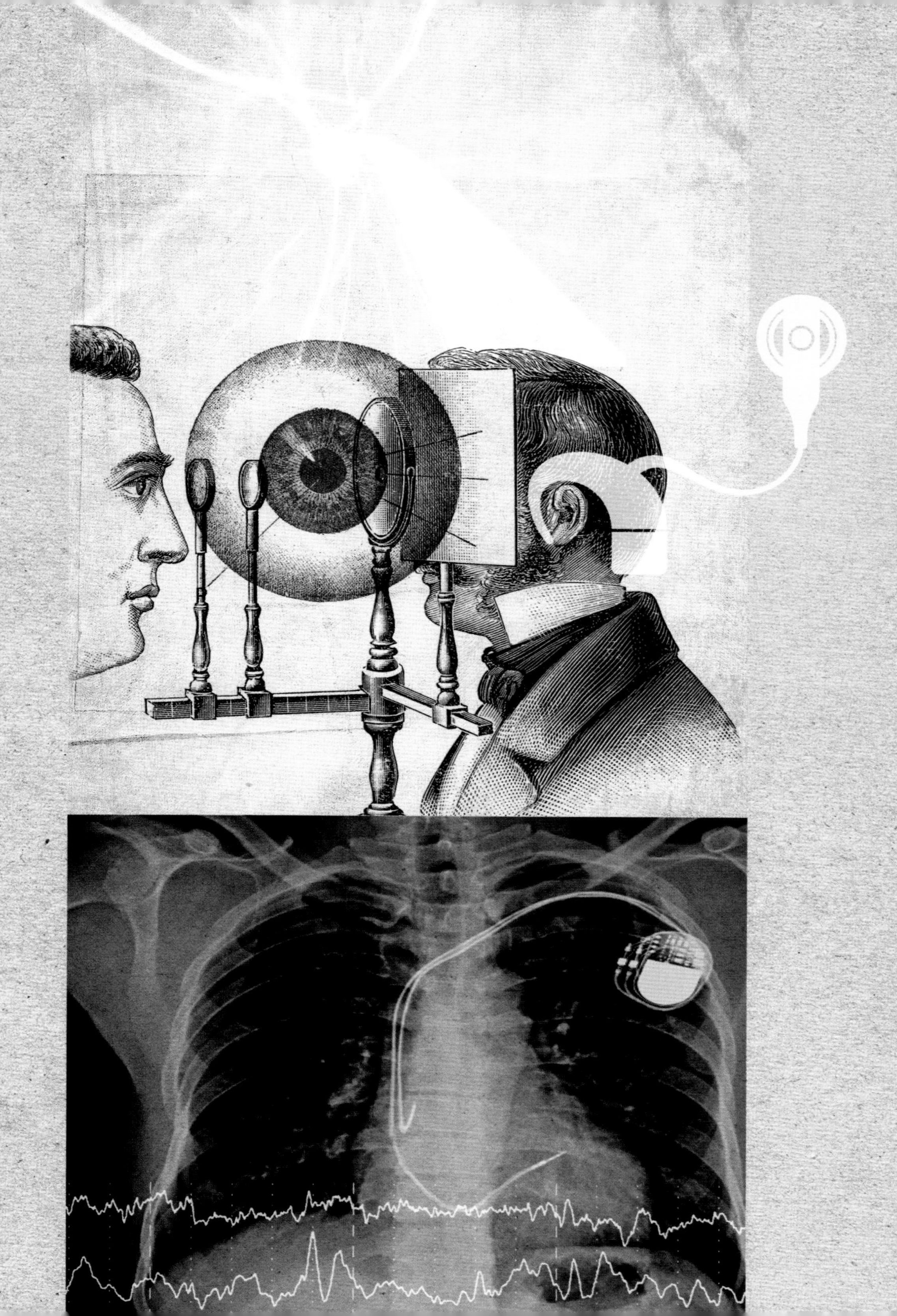

AEROSPACE & TRANSPORT ENGINEERING

AEROSPACE & TRANSPORT ENGINEERING
GLOSSARY

aerodynamics Principles governing motion of aircraft, particularly forces arising from the flow of air around moving aircraft.

aileron Small, movable sections near outermost parts of wings on an aircraft, enabling the roll angle of the aircraft to be adjusted, in order to make a stable turn. Some ships have ailerons, also called 'horizontal stabilizers', projecting from the hull, to stabilize the ship and reduce rolling motion caused by waves.

alloy Mixture of metals and other elements to improve material properties. Soft aluminium becomes strong like steel when alloyed with zinc, magnesium, copper and other elements.

bearing Machine component that allows another component to rotate or slide with minimal friction; for example, a ball or roller bearing.

carbon fibre Extremely high-strength fibre made from pure carbon; used as reinforcing fibre in polymers to produce very stiff, high-strength aircraft parts.

ceramic Hard, usually brittle, non-metallic materials, often used in high-temperature applications, or where excellent insulation is needed.

composite material Material comprising two or more distinct materials with complementary properties; for example, carbon fibre in a polymer matrix.

creep Slow extension under stress, particularly affecting metal alloys at high temperatures in aircraft engines.

crosstie Wood, steel or concrete support for rails.

delamination Separation of stiffening fibres from the surrounding matrix in composite materials.

drag Force impeding motion of a body through a fluid.

elevator Small, movable section of horizontal tail surface of aircraft that creates variable lift force to enable the angle of the aircraft relative to the air stream to be adjusted, enabling the lift force from the wings to be controlled.

fatigue Progressive failure of metal components subjected to repeated cyclical load, for example aircraft wings. Engineers have to allow for a lower maximum stress to avoid fatigue failure in these components.

gradient Slope, expressed as change in elevation along a given distance.

machine learning Computer program designed to be able to improve its behaviour with the accumulation of data representing past behaviour and performance.

magnetic levitation Supporting a structure using magnetic forces – active electronic controls are usually needed to achieve stable operation.

model Set of mathematical equations, usually embodied in a computer program, which engineers use to predict the behaviour of an engineering system. Engineers also use physical models, often scale models with identical proportions but a different size relative to the system being studied.

reliability Time performance of machinery or plant with no failures, usually measured as mean time between failures (MTBF).

rudder Small, movable vertical surface at rear of a ship or aircraft, enabling turning of a ship or aircraft.

sloshing Tendency of large waves to form in liquid stored in mobile tanks, creating large impact forces that can destabilize a vehicle or ship.

suction pile Foundation or anchor point consisting of a large steel tube capped at the top and driven into the seabed or riverbed by extracting water.

thrust Propulsion force developed by aircraft engines, rocket motors and propellers on ships.

turbulent flow Erratic fluid flow with eddies and rapid small random variations, typical of high-speed fluid flow.

viaduct Long bridge to enable a road or railway to cross a river or uneven ground.

RAILWAY ENGINEERING

30-second foundation

3-SECOND CORE
Raiways enable goods and people to travel fast over long distances because the tracks are planned for very low rolling resistance. High-speed trains can move people between cities faster than air travel.

3-MINUTE IDEA
Railways were invented for moving heavy loads in mines – literally, roads made from rails. George and Robert Stephenson adapted this idea to move people and goods between English cities in the nineteenth century. Soon railways were built in Europe, America and other countries. In Europe, tracks were planned for low rolling resistance from the start. However, in America distances were longer and capital was harder to raise, so tracks were improved later, as money from operations permitted.

Often, an engineer's job is to improve existing technology. Railways move heavy goods and people quickly over long distances with less energy than road vehicles. Track design influences the pulling force needed from the locomotive: rolling resistance. Rolling resistance is proportional to uphill gradient through mountains and valleys and increases on curves. Engineers can improve a railway by realigning the track: instead of going down into valleys, high bridges and viaducts can keep the track on a level path; tunnels save long zigzag climbs up mountains. More gradual curves also reduce rolling resistance. While expensive and often difficult to build, viaducts, tunnels and other improvements allow smaller locomotives to move heavier trains faster on more direct routes. Low-friction bearings on locomotives and rolling stock can also reduce rolling resistance, as can supporting the track on concrete crossties or even continuous reinforced concrete foundations. High-speed passenger rail networks in Japan, France, Germany and China demonstrate the value of these improvements. Instead of using older rail lines, often completely new lines have been constructed to minimize rolling resistance. These lines enable trains to move at speeds that make them competitive with air travel.

RELATED TOPICS
See also
CIVIL ENGINEERING
page 34

GEOTECHNICAL ENGINEERING
page 40

MECHANICAL ENGINEERING
page 56

3-SECOND BIOGRAPHIES
GEORGE STEPHENSON
1781–1848
English civil and mechanical engineer, considered to be the 'father of railways'; developed the first inter-city passenger railways in England.

ROBERT STEPHENSON
1803–59
English civil and railway engineer, son of George, who pioneered locomotive design and railway and bridge building.

30-SECOND TEXT
John Blake

Railways are as straight and level as possible to enable high speed with minimum energy.

1972
Born in Jilin province, China

1991
Accepted by Shanghai Railway University (now Tonji University)

1995
Graduates in Electric Drives and Control System Engineering. Joins CRRC Qingdao Sifang Co. Ltd.

2006
Appointed as Directing Designer for 300 km/h (185 mph) trains

2008
Appointed as Directing Designer for first Chinese-designed high-speed train, CRH380A

2010
CRH380A prototype achieves world record speed of 486.1 km/h (302 mph)

2012
Appointed Vice General Manager and Chief Engineer, CRRC Qingdao Sifang Co. Ltd

2013
Commences design work on CR400AF train, specifically optimization of the electric motor drive system. Energy efficiency, safety and reduced noise are also important objectives.

2015
Leads design of CRH2G, operating at low temperatures and in sandy desert regions

2017
CR400AF train enters commercial service

LIANG JIANYING

Trains have been the focus of Liang Jianying's life, from her earliest memories living close to the local station in a small town in mineral-rich Jilin province in northeast China. Liang loved watching passing trains and admired their creators. Her love for trains was tested when returning home from university in Shanghai for the 1992 spring festival, sitting for more than 50 hours in a carriage packed with people, many on the floor, and enduring motion sickness. Liang resolved to do her best to prevent people suffering like that in otherwise happy festivities. After graduating in her chosen discipline, Electric Drives and Control Systems Engineering, at Shanghai Railway University in 1995, she joined Qingdao Sifang, a leading Chinese train manufacturing company.

In 2004 the design team of CRRC Qingdao Sifang Co. Ltd built the first of China's 200 km/h (125 mph) high-speed trains with the help of advanced technology from other countries. By 2006 Liang was the directing designer for trains running at 300 km/h (185 mph). Her determination helped her endure strenuous working hours, leaving home whilst her daughter was asleep and returning after her bedtime. Imported technology from other countries still lay at the core of the design, but Liang's team was determined to develop Chinese designs.

By 2008 Liang was leading China's first effort to design its own high-speed train, the CRH380A, which set the world speed record for a commercial train running on rails at 486.1 km/h (302 mph). She built on the results of this effort to lead design of the faster CR400AF, with the aim of exporting China's own high-speed rail technology internationally. With a continuous operation speed of 350 km/h (217 mph), this train was tested to speeds of 420 km/h (260 mph). After more than 600,000 km (370,000 miles) of test running, it entered commercial service in 2017.

Now Liang is focused on still higher speeds using magnetic levitation, aiming for 600 km/h (370 mph). Asked about her greatest achievements so far, she points to the speed record set by the CRH380A, and the fastest speed achieved by two CR400AF and CR400BF trains safely passing each other in opposite directions, 420 km/h (260 mph), in July 2016. She is also proud of the energy efficiency her team has achieved, consuming less than 4 kWh per passenger per 100 km (60 miles).

Zhigang Ji & James Trevelyan

FLOATING FACTORIES

30-second foundation

More and more minerals, oil and gas are coming from under the sea. Huge floating factories are being built to extract and process these resources. Building refineries on ships requires many new technologies. Shell's huge floating liquefied natural gas (FLNG) plant, Prelude, reveals some of these challenges. Nearly 500 m (1,650 ft) long, it is one of the largest floating structures in the world, and has been designed to withstand 400 km/h (250 mph) winds in category 5 hurricanes. Over 500 km (300 miles) from the nearest land-based support, the ship has to operate for 25 years before returning to a dry dock for overhaul. Space and motion pose more challenges. Engineers had to design the entire process plant to operate safely and reliably on a heaving ship in one quarter of the area needed for an equivalent land-based plant. Liquified gas is stored in vast insulated tanks designed to curb sloshing that could otherwise destabilize the vessel. Special articulated arms enable liquid gas at -160°C (-250°F) to be safely offloaded to tankers, even in rough seas. A worldwide collaboration involving thousands of engineers was needed, designing and constructing the components to the highest quality and reliability standards, and assembling them all at Geoje Island, Korea.

3-SECOND CORE

Floating factories enable seabed resources to be extracted far from land. Storms, ship motion and the paramount need for safety pose special engineering challenges.

3-MINUTE IDEA

Safety is paramount, particularly for FLNG, not only for the people on board but also for the marine environment, economics and reputation. Companies have learned that compromising safety for production economies causes catastrophes that can bring even multinational companies to their knees. Far from land-based support, a fire or explosion has to be prevented, or, in the worst case, contained and extinguished. Safety has to be designed in from the start and no defects can be tolerated, even in the paint work.

RELATED TOPICS

See also
CHEMICAL ENGINEERING
page 76

ORGANIZATIONAL SAFETY
page 82

PROCESS PLANT SAFETY
page 84

3-SECOND BIOGRAPHIES

ZHENG HE
1371–ca. 1434
Pioneered long-distance navigation and established trading colonies at several Indian Ocean ports.

ISAMBARD KINGDOM BRUNEL
1806–59
Developed large metal steam-powered ships, railways and spectacular bridges.

ROBERT BEA
1937–
Investigated major offshore disasters, guiding the design of much safer floating structures.

30-SECOND TEXT

James Trevelyan

Huge floating factories will extract and process minerals and food from the oceans.

FUNDAMENTALS OF AERODYNAMICS

30-second foundation

Aerodynamics began with Leonardo da Vinci's 'ornithopter' invention in 1485. However, achieving sustained flight of heavier-than-air machines took until 1903, when the Wright brothers made their first flight near Kitty Hawk, North Carolina. Aerodynamics explains how fluids such as air move around objects, at least approximately. It explains aircraft flight, as well as trailer trucks, race cars, hydrofoil racing boats and even curved baseball throws. By using aerodynamics, engineers can calculate forces and moments on the object from the flow field: the pattern of fluid motion. Flow fields describe velocity, pressure, density and temperature, which vary with position and time, and also depend on properties such as shape and fluid viscosity. Flow fields can be measured in wind tunnels or computed from equations derived from knowing that mass, momentum and energy have to be conserved. For flight, the four key forces are lift, drag, thrust and weight. Lift and drag arise from the flow field. Lift has to overcome weight for an aeroplane to fly, and thrust from engines has to overcome aerodynamic drag, which tends to slow the plane. Smaller forces from elevators and a rudder stabilize aeroplane orientation and provide turns when needed.

3-SECOND CORE

Aerodynamic principles help engineers design aeroplanes, cars, ships and trains, enabling them to predict airflow and the resultant forces. The same principles help explain how birds and insects fly.

3-MINUTE IDEA

Aeroplanes fly because they are able to generate lift from the speed of air flowing over and under the curved wing surfaces. Air moves faster above the wing than below, so pressure is less above the wing, generating an upwards force. Forward motion results from engine thrust or wing movements for birds. Stable flight depends on maintaining correct orientation relative to the airflow. Today, computers help pilots control planes for added safety and reliability.

RELATED TOPICS

See also
MECHANICAL ENGINEERING
page 56

AEROSPACE MATERIALS
page 124

LESSONS FROM SPACE
page 126

3-SECOND BIOGRAPHIES

LEONHARD EULER
1707–93
Laid many foundations for modern mathematics and aerodynamics.

WILBUR & ORVILLE WRIGHT
1867–1912 & 1871–1948
Pioneers credited with inventing, building and flying the world's first successful aeroplane.

ANDREI TUPOLEV
1888–1972
Developed over 100 different aircraft types, despite being imprisoned and closely guarded during Stalin's era.

30-SECOND TEXT

George Catalano

Engineers predict motion of flying objects using aerodynamics.

AEROSPACE MATERIALS

30-second foundation

Since the hot air balloons of the eighteenth century, aircraft engineers have worked at the extreme limits of material performance. Modern aeroplanes demand extreme reliability in extreme temperatures. Passenger jets cruise in air temperatures of around -50°C (-58°F), while some engines endure more than 1,000°C (1,800°F). Nickel superalloys coated with a ceramic thermal barrier are used in the combustion chamber and turbine blades because they resist creep at high temperatures. Materials subjected to varying loads can develop invisible cracks that slowly extend each time the load peaks. When the crack is large enough, a part like a turbine blade can suddenly fracture – a fatigue failure. Engineers ensure that the peak loads are within known fatigue limits to avoid these failures. In terms of strength, weight and stiffness, carbon-fibre reinforced polymers are superior to aluminium alloys, which are superior to steel. However, composites can be susceptible to impact damage and delamination. Though heavier, some high-strength titanium or steel alloys are used in critical components. Engineers use computers to predict the behaviour of aeroplane and spacecraft structures, but only full-scale experimental testing can guarantee optimal performance and durability.

3-SECOND CORE
Aircraft and spacecraft require special materials that are engineered to be light, stiff, strong, durable and resistant to harsh environments. Experiments and computer simulations ensure safe design.

3-MINUTE IDEA
Carbon-fibre composites combine the advantages of the fibre and a rigid polymer into a single material. The fibres are extremely strong along the fibre axis. Engineers design layered composites to resist loads from multiple directions. Shells formed from composite layers sandwiched around honeycomb cores have increased stiffness. New directions for composites include the development of high-temperature polymers and multi-functional materials that may enable the elimination of redundant materials.

RELATED TOPICS
See also
MATERIALS ENGINEERING
page 60

FUNDMENTALS OF AERODYNAMICS
page 122

LESSONS FROM SPACE
page 126

3-SECOND BIOGRAPHIES

SIR JAMES ALFRED EWING
1855–1935
Investigated magnetic materials and connected metal fatigue failure with microscopic defects in metal crystals.

LEONARD BESSEMER PFEIL
1898–1969
Developed one of the first nickel alloys that could withstand the requirements for jet engine turbines.

AKIO SHINDO
1926–2016
Produced sufficiently high-strength carbon fibre to manufacture aircraft parts.

30-SECOND TEXT
Matthew L. Smith

Strength, stiffness and lightness are crucial materials attributes.

LESSONS FROM SPACE

30-second foundation

Outside Earth's atmosphere, spacecraft are subjected to vacuum conditions, temperature extremes, radiation damage and the threat of impacts from micro-meteoroids or space debris. Spacecraft must traverse debris fields around the Earth, and space probes may also encounter dust and toxic gases in planetary atmospheres. The high radiation experienced in space is a threat to both humans and electronic equipment. Spacecraft routinely experience ultra-wide-ranging temperatures, from as low as -230°C (-382°F) to 200°C (392°F) or higher. Batteries and electronics are vulnerable to such extreme temperatures. Mitigating the effects of the extremes of space is an engineering challenge. Engineers develop thermal and structural computer models of the spacecraft, then test and certify prototypes in conditions simulating space. Vibration tables and acoustic chambers simulate launch conditions. Vacuum chambers test resilience to low pressures and extreme temperatures. Electronics are redesigned with shielding to resist the damaging effects of radiation and cosmic rays. Spacecraft have ventured into the extreme environments of space, but a major lesson learned from accidents is that success must not breed complacency: the dangers from the extremes of space can be temporally overcome, but never mastered.

3-SECOND CORE

Spacecraft experience an environment of extremes of temperature, pressure, shock and radiation. Engineering designs have overcome these challenges, opening up space for exploration and commercial use.

3-MINUTE IDEA

Today, private ventures are making access to space more convenient, reliable and routine. These companies design, manufacture and launch advanced engines, rockets and spacecraft. Driving the engineering are requirements for low recurring cost, reusability, high performance and reliability. Significant achievements so far include the return landing of a rocket, reuse of an orbital-class rocket and the first cargo delivery to the international space station by commercial spacecraft.

RELATED TOPICS

See also
MECHANICAL ENGINEERING
page 56

ORGANIZATIONAL SAFETY
page 82

AEROSPACE MATERIALS
page 124

3-SECOND BIOGRAPHIES

ROBERT HUTCHINGS GODDARD
1882–1945
Considered to be the inventor of modern liquid-fuelled rocket propulsion.

WERNHER VON BRAUN
1912–77
Led rocket launcher development in Germany and later America.

MARY WINSTON JACKSON
1921–2005
First African-American engineer in NASA.

30-SECOND TEXT

John Krupczak

Even in launch, rockets experience powerful shaking along with intense sound waves.

DRIVERLESS CARS

30-second foundation

3-SECOND CORE

Driverless cars use software and on-board sensors to understand their surroundings and plan movement. Conceptual simplicity is challenging to implement, and it will bring profound social changes.

3-MINUTE IDEA

Commercial driverless cars will not be perfect: economic engineering requires compromises. Because they are driven by software and algorithms, learnings following an accident can be shared as software updates across entire fleets of vehicles. While no accident is acceptable, it does mean that every accident – even a near miss – can help reduce the chances of a repeat. Human drivers rarely share experience like this. Could this capacity make machines the best drivers?

Instead of human drivers, driverless cars have computers, lasers, radar, cameras and GPS. The computer has to answer three questions: Where am I? What surrounds me? What should I do? Special software compares incoming sensor data with memorized 3D maps to figure out where the car is within a few centimetres. The processing of camera images and laser data provides information about moving and static obstacles on or near the road around the car. Machine learning and computer vision help computers categorize nearby obstacles. For example, cars move differently to bikes and pedestrians, so the computer can identify the type of obstacle and then predict what is likely to happen over the next few seconds. The computer makes a plan on how to drive for the next few seconds, so the car will avoid obstacles and provide the occupant with a smooth and natural ride. The computer then adjusts the accelerator or brakes and operates the steering wheel. The computer interprets a new set of sensor measurements and recalculates the plan again in a continuously repeating cycle many times every second. Engineers draw on computer science, robotics, machine learning, sensing, optimization and mathematics principles to design these systems.

RELATED TOPICS

See also
MECHATRONICS
page 62

ROBOTICS & AUTOMATION
page 70

SOFTWARE ENGINEERING
page 100

3-SECOND BIOGRAPHIES

HENRY FORD
1863–1947
Pioneered factory production lines in 1913: his famous Model-T car accounted for half of America's cars by 1918.

SOICHIRO HONDA
1906–91
Founded the Honda motor company; Japan's leading automotive engineer of the twentieth century.

RUDOLF E. KÁLMÁN
1930–2016
Laid the foundation of modern navigation systems.

30-SECOND TEXT

Paul Newman

Driverless vehicles are poised to transform our lives and bring economic and safety improvements.

ENGINEERING THE FUTURE

ENGINEERING THE FUTURE
GLOSSARY

alumina Aluminium oxide produced as an intermediate product in aluminium production. Alumina powder is produced near bauxite mines (most common source of aluminium) and transported to aluminium smelters, which require large amounts of low-cost electricity.

anaerobic (chemical engineering) Process that takes place in the absence of oxygen, particularly using bacteria that thrive in an environment with little or no oxygen present.

autonomous machine Machine with sufficient sensing, computing and communication capacity to operate automatically, without a human controller being in charge.

demand (electrical engineering) Requirement for electricity to be provided, determined by the number and power requirements of all devices currently connected and switched on in an electricity supply network.

demand management (electrical engineering) When demand is high, grid controllers can request major power users to switch off or reduce their power consumption in return for charging less for power consumed. Many devices can reduce their power consumption for a while without noticeable effects.

diode (electronics) Electronic component that allows current flow in only one direction.

drone Autonomous or remotely operated vehicle, commonly a small helicopter.

GPS Global positioning system – an arrangement of satellites and radio communication that enable a suitably equipped receiver to calculate its position on Earth with great accuracy.

green chemistry Development of chemical processes that reduce or eliminate harmful and hazardous substances.

grid (electrical engineering) Entire electric power supply system including electricity generating stations, transformers, switch yards and powerline interconnections.

industrial ecology Term used for two or more mutually dependent industries that use waste from other industries as process inputs, avoiding the need to discharge waste into the environment.

LiDAR Light detection and ranging using a laser beam; use of triangulation or timer to detect the distance of an object reflecting the beam. The direction of the beam indicates the direction of the object.

light emitting diode Diode that emits coloured light when current flows. Modern LEDs are energy-efficient light sources with many applications, including car headlamps and medical devices. Unlike earlier light sources, they release relatively little heat and so stay cool.

power generation plant Electricity-generating station that can run on fossil fuels such as coal, oil or natural gas, or use heat from a nuclear reactor or solar collectors.

radar Transmission of high-frequency radio signals and detection of echoes: the time taken for the echo to come back indicates the distance of the reflecting object; the direction of the beam indicates the direction of the object.

sensor Device that can measure a physical property and generate a signal indicating the measured value. For example, a thermocouple measures temperature and generates a small electric voltage indicating the temperature.

slag Waste from iron and steel making plants.

smart machine Generic term used for a machine with built-in sensing, information and communication capability that enables data exchange with other devices and systems; possibly capable of autonomous actions without human interaction.

sonar Transmission of high-frequency sound and detection of echoes: the time taken for the echo to come back indicates the distance of the reflecting object; the direction of the beam indicates the approximate direction of the object.

WiFi Radio communication technology developed to provide convenient wireless Internet connections for mobile devices such as laptop computers and phones.

THINKING DIFFERENTLY

30-second foundation

Until this century, engineers designed for an infinite world. Water was always available and waste could be freely discharged. There were rules against pollution, but enforcement was weak. Now there are powerful financial deterrents. Today, every large project will confront limits on water supply and discharge, and countless other restrictions. Banks' reputations can be damaged by projects they finance. When the world's media hear about toxic waste being discharged, banks may recall loans. Sensitive measuring instruments are widely available from online shops, leading to leaks and pollution often being discovered by amateur scientists or community activists, who can close down whole projects. That's great news for engineers working on recycling and waste recovery projects. Refineries running on renewable energy, which clean air and water rather than polluting them, recover valuable waste, and are so quiet and clean people want to live next to them – these are the projects that attract cheaper long-term finance. Even in the developing world, companies can secure funds from wealthy countries to improve waste management and reduce emissions.

3-SECOND CORE
Banks' reputations often depend on the environmental performance of projects they finance, empowering engineers to design clean, green factories and refineries running on renewable energy.

3-MINUTE IDEA
Eliminating greenhouse gas emissions can only be achieved with clever engineering and smart energy management. New technologies take 25 years or more to mature, so the challenge for engineers is to devise simple combinations of existing technologies that capture public imagination and achieve significant cost savings to secure the finance needed for large-scale investments.

RELATED TOPICS
See also
ENVIRONMENTAL ENGINEERING
page 50

ENERGY & FINANCE
page 140

3-SECOND BIOGRAPHIES
JAMES LOVELOCK
1919–
Proposed the Gaia hypothesis, devised methods for detecting life on other planets and first identified atmospheric accumulation of CFC gases.

JOHN GRILL
1945–
Founded WorleyParsons and recognized how sustainability affects all major engineering projects, particularly supply and disposal of water.

30-SECOND TEXT
James Trevelyan

Clean, green, sustainable engineering is the future.

INNOVATION

30-second foundation

3-SECOND CORE

Engineering innovation transforms ideas into real solutions. Everything around us has come from engineering innovations: housing, transport, food, clothing, energy, water supply and sanitation, computers, communication and health.

3-MINUTE IDEA

Engineering innovations have facilitated social justice, economic empowerment and political revolutions: the Arab Spring relied on Facebook and Twitter. Innovation enables economic growth by increasing productivity. Low-cost mobile phones helped create large profitable businesses, displacing expensive, inefficient government monopolies, and providing affordable global communications and universal Internet access.

Innovation and invention is at the core of engineering. The word 'engineer' is derived from the Latin words *ingeniare*, meaning 'to contrive, devise', and *ingenium*, meaning 'cleverness'. Most inventions need a collaboration with engineers, marketing experts, manufacturers and investors to convert an idea into a business distributing new products to people who need them. Engineers create a succession of prototypes, first in a laboratory, later in the field, learning from testing and customer evaluations, gradually improving performance, building confidence and evolving product designs. New spaceships, mobile apps, materials, processes and engines – they all follow the same process, and it takes months or years. Inventors register patents in each major country where the products will be used, providing investors with exclusive rights to sell the products for up to 25 years. Engineer Shuji Nakamura and his colleagues researched blue light emitting diodes (LEDs) for decades. Today, billions of blue LEDs, with red and green cousins, provide energy-efficient lighting and colour displays on smart phones and televisions. Innovations like this help meet human needs for light and information with much less materials, energy and pollution, helping to build a sustainable world for everyone.

RELATED TOPICS

See also
ELECTRONIC ENGINEERING
page 96

SOFTWARE ENGINEERING
page 100

INFORMATION & TELECOMS
page 106

3-SECOND BIOGRAPHIES

CHARLES F. KETTERING
1876–1958
Led development of countless automotive innovations, such as electric starting motors and leaded gasoline.

JOHN O'SULLIVAN
1947–
Invented WiFi technology that provides local Internet access for computers and phones.

SHUJI NAKAMURA
1954–
Awarded Nobel physics prize for invention of efficient blue and white LEDs.

30-SECOND TEXT

Marlene Kanga

Engineering innovation has been transforming our world for millennia.

4 January 1936
Born in Niederkassel, Germany

1961
Graduates in aeronautical engineering at RWTH Aachen. Joins German Aerospace Research Establishment, DLR.

1965
Gains master's degree in control engineering at Princeton University, USA

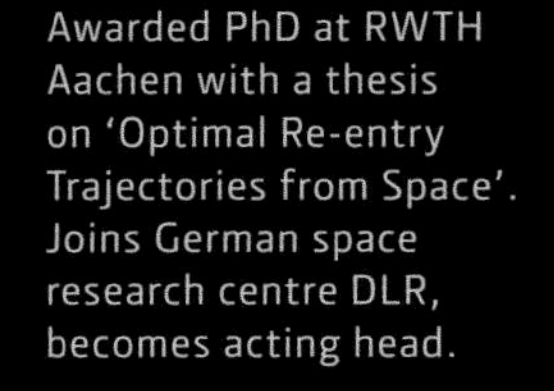

1969
Awarded PhD at RWTH Aachen with a thesis on 'Optimal Re-entry Trajectories from Space'. Joins German space research centre DLR, becomes acting head.

1975
Joins Bundeswehr University, Munich, seeking more time for research

1977–82
Researches computer vision guidance for aircraft, helicopters, spacecraft and ground vehicles

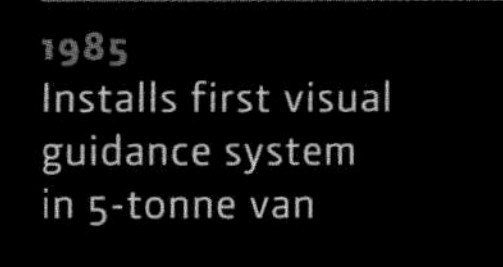

1985
Installs first visual guidance system in 5-tonne van

1987
Demonstrates high-speed autobahn driving

1987–1995
Leads the Eureka Prometheus project (PROgraMme for a European Traffic of Highest Efficiency and Unprecedented Safety)

1995
Achieves substantial autobahn journeys and automatic driving at high speed, with occasional manual intervention

2001
Completes third-generation visual guidance system and demonstrates automatic driving on networks of minor roads. Retires from teaching at Bundeswehr University.

ERNST DIETER DICKMANNS

Twelve years after he was born in a village near Cologne in 1936, the son of a teacher, Ernst Dickmanns was driving tractors around in wartime Germany. Studying calculus from aged 15, his driving experience helped him to see differential equations as an ideal tool for motion control, and inspired him to study advanced aeronautical engineering and flight control – although it would not be until much later in his career that he turned his attention to driverless cars.

Dickmanns' initial passion was aerospace engineering, and so he studied aerospace and aeronautics at RWTH Aachen, followed by control engineering at Princeton University. On his return to Germany, he worked for the German space research centre, developing flight dynamics and trajectory optimization, and then segued into satellite control. By 1977, Dickmanns had started looking into providing 'vision' for computers on vehicles. Vision-guided vehicles had first appeared in 1970, though few moved faster than a mile per hour, and engineers assumed that driverless cars would always need radio guidance from cables buried under roads. However, in 1987, a fully automatic, vision-guided 5-ton van, created by Dickmanns and his team at the Munich Bundeswehr University, reached speeds of 60 mph (100 km/h).

This breakthrough came from Dickmanns' realization that the differential equations used in aircraft navigation and autopilots could be adapted to control cars, if visual guidance from cameras was fast enough – at least ten times a second. Collecting images was easy: analysing the images at that rate was impossible. Dickmanns realized that analysing small windows within the images would be much faster, and his colleague Volke Graefe made it work. They used two cameras, one with a wide viewing angle to follow the nearby edges of traffic lanes and one with a telephoto lens to measure the road curvature ahead. The seven-year, €749,000,000 Eureka Prometheus project was approved, with Dickmanns in a leading role.

By 1995, Dickmanns' team demonstrated 1,500-km (1,000-mile) journeys along crowded autobahns, with fully automatic driving at speeds of up to 175 km/h (100 mph). The cars could change lanes automatically to overtake traffic and required only occasional manual intervention. Even with these demonstrations, it took years for Dickmanns' ideas to gain acceptance amongst computer science and artificial intelligence communities.

By the time of his retirement in 2001, Dickmanns' team had overcome many challenges. His ideas have been embodied in today's driver-assist technologies, and engineers are currently testing prototype driverless cars on city streets, whilst legal and regulatory issues are settled in anticipation of widespread adoption.

James Trevelyan

ENERGY & FINANCE

30-second foundation

3-SECOND CORE
Mobile phone systems enable people to buy on credit, such as solar electricity with battery storage. Like pre-paid mobiles, people pay as they go.

3-MINUTE IDEA
The fossil fuel age is giving way to an energy revolution based on renewables and smart energy storage coupled with market places driven by intelligent machines and variable speed processes that adapt to energy availability and price. Telecommunication systems that act as trust brokers secure financial transactions that used to rely on fragile human relationships with banks as intermediaries. Together, they have the potential to transform life for most people on the planet.

Twentieth-century industries relied on huge power stations, national electricity grids and centralized control. Centralized banks raised finance to build these systems, run by engineers and bureaucrats. Elaborate and disciplined policing ensured sufficient compliance by people to pay taxes and bills. Yet resource consumption is unsustainable and pollution threatens the global climate. A financial revolution is bringing renewable energy and other services to the Third World without the same need for engineers, bureaucrats and policing. Mobile phones, more than a means of communication, verify identity and provide secure, trustworthy payment systems and tax collection without requiring bank accounts. Using what is known as mobile fintech, farmers will pay for refrigerated storage as they use it, with credit secured by crop production. Already, this technology has enabled cash-strapped farmers to lease solar panels with battery storage and earn extra income selling electricity they do not use. Suppliers are able to extend credit knowing that their customers have to keep up with payments to use the machines. Mobile fintech enables industrial processes to run on variable renewable power supplies, buying the cheapest energy whenever it is available.

RELATED TOPICS
See also
POWER GENERATION & ENERGY STORAGE
page 78

INFORMATION & TELECOMS
page 106

THINKING DIFFERENTLY
page 134

3-SECOND BIOGRAPHIES

RUSSELL SHOEMAKER OHL
1898–1987
Developed semiconductor diodes and the first silicon solar cells, which led to the development of transistors.

JOHN GOODENOUGH
1922–
Contributed developments needed for high-capacity rechargeable lithium batteries.

MOHAMMED 'MO' IBRAHIM
1946–
Led development of mobile phone networks in Africa.

30-SECOND TEXT
James Trevelyan

Mobile Internet communications enable even small traders to access global markets.

RESOURCE SCARCITY

30-second foundation

A growing population combined with wasteful material usage has created scarcities. The World Bank has forecast that global waste generation by 2025 will exceed 6 million tonnes a day, 70 per cent more than in 2015. Learning to see waste as a valuable resource can solve many problems. For example, recovering copper from electronic waste is sustainable, but also, the copper is far more concentrated than that in metal ores. Discarded printed circuit boards contain 10 to 20 times more copper metal than the same weight of copper ore. Electronic waste also contains precious metals such as gold and silver. The gold content can be more than 20 times greater than natural gold-bearing rock, creating a convincing economic justification for recycling. Conventional recycling converts like for like, using glass or plastics to make more of the same; however, special processing is needed for other waste streams. Engineers are working on 'micro-factories' – modular production facilities that recover resources from waste, converting materials back into their original composition and structure or into new materials. Micro-factories will be built where waste accumulates, reducing transportation costs and creating local employment.

3-SECOND CORE

Extracting valuable materials from waste will gradually become more profitable than mining for increasingly scarce natural resources. Adaptive transformation processes are needed as waste composition changes continuously.

3-MINUTE IDEA

Micro-factories will need to adapt to changing waste composition. An electronics waste stream may have more laptops one day, hard drives another. Robots and scanners using vast data sets will collaborate with people who may still be learning – the scanners can help people identify items like radioactive fire alarm components that need special handling. People can more easily disassemble products that would be too complex for robots to handle.

RELATED TOPICS

See also
PLASTICS & FERTILIZERS
page 86

THINKING DIFFERENTLY
page 134

CONTROLLING POLLUTION
page 148

3-SECOND BIOGRAPHIES

MAHATMA GANDHI
1869–1948
Anticipated resource shortages and advocated alternative production approaches.

CARL SHIPP MARVEL
1894–1988
Developed a synthetic rubber manufacturing process and high-temperature polymers for aerospace applications.

AMORY LOVINS
1947–
Advocates environmentally friendly developments in energy and materials.

30-SECOND TEXT

Veena Sahajwalla

Engineers are working towards a future where most products are made from reused materials.

FEEDING OUR WORLD

30-second foundation

3-SECOND CORE
Food engineering developments, particularly in less developed countries, are crucial for our common future on Earth. Food engineers enable affordable, consistent, year-round food supplies.

3-MINUTE IDEA
Food engineering brings together plant and animal breeders, producers, processors, food scientists, transporters and supermarkets to provide consistent supplies. Plant varieties provide extended harvest seasons and consistent ripening – with size, consistency and shape adapted for mechanized handling and processing. Packaging protects food from damage and extends shelf life. This complexity enables affordable year-round stable food supplies.

Food engineering is a vital part of our common future. One third of food grown on Earth is wasted. However, in many less developed countries (LDCs), where hunger is common, waste is far more prevalent because of storage, processing and distribution losses. More could be grown with better education and infrastructure. While plastic containers and wrapping cause a global waste headache, they keep food fresh, enormously reducing wastage. Refrigerated stores could eliminate most food wastage in developing countries, and appropriate packaging could prevent much of the rest. The planet's increasing population could be fed without growing more. A major obstacle is access to credit to buy the necessary food processing and storage machinery. Newly emerging secure electronic payment systems linked with mobile phone-enabled machines will enable machinery suppliers to extend credit to small enterprises. The user has to keep up payments through their mobile phones for the machines to operate. These technologies will enable farmers and rural enterprises to buy food processing and storage machinery on credit, without needing bank accounts or land to secure loans. New biodegradable food packaging materials can solve the plastic waste problem.

RELATED TOPICS
See also
PLASTICS & FERTILIZERS
page 86

ENERGY & FINANCE
page 140

WATER SECURITY
page 146

3-SECOND BIOGRAPHIES
BRYAN DONKIN
1768–1855
Pioneered production of processed food in tin cans.

CLARENCE FRANK BIRDSEYE II
1886–1956
Noticed how fish could be snap-frozen to retain their food qualities when thawed later, and founded the frozen food industry.

G. HOWARD KRAFT
1908–83
Pioneered inert gas-filled packaging to extend the shelf life of cheese and other foods.

30-SECOND TEXT
James Trevelyan

Food engineering, like water engineering, is crucial for our future.

WATER SECURITY

30-second foundation

Water is essential for life, and there is no substitute. The UN's 2030 Agenda set the target of ensuring availability and sustainable management of water and sanitation for all. In 2015, 844 million people lacked a basic water service; 2.1 billion lacked safely managed drinking water; 4.5 billion lacked a safely managed sanitation service; more than 2 billion lived in countries with acute water shortages. Adapting to climate change is making the challenge even more difficult. Engineers are needed to design and operate dams and reservoirs, channels, pipelines, water treatment plants and also for planning and managing water resources. New nature-based engineering solutions are emerging to improve rivers, underground aquifers and urban drainage. Safe wastewater reuse and desalination will also be needed, along with new irrigation technologies, as agriculture demands account for 70 per cent of all water use. Engineers also help prepare for floods and droughts, reducing economic losses from natural disasters. Integrating information technology into water and sanitation systems offers exciting new solutions. Improved tracking of water flow and consumption can provide the improved security needed to attract private investment while still ensuring that reliable services meet basic human needs.

3-SECOND CORE

Engineers need new water and sanitation technologies to meet recently updated sustainable development goals, providing safe water and sanitation for all.

3-MINUTE IDEA

Water is critical for drinking, food production and to provide cooling for electricity production and industrial processes. Water engineers have to work with farmers, government regulators, energy producers and process engineers. Water conservation solutions require an understanding of human behaviour and how to influence farmers, industrial users and all other users to make wise use of limited resources. Integrating knowledge of human behaviour into engineering solutions requires close collaboration with social scientists.

RELATED TOPICS

See also
CIVIL ENGINEERING
page 34

TAMING GREAT RIVERS
page 46

ENVIRONMENTAL ENGINEERING
page 50

3-SECOND BIOGRAPHIES

MANUEL LORENZO PARDO
1881–1953
Spanish civil engineer, founding director of first river basin organization in the world; helped to transform Spain in the mid-twentieth century.

GIOVANNI LOMBARDI
1926–2017
Swiss civil engineer who founded the Lombardi firm, renowned for tunnel and dam construction; widely respected for sharing his knowledge with the civil engineering community.

30-SECOND TEXT

Tomás A. Sancho

Engineers' contributions are essential to reach the UN's target of 'water for all'.

CONTROLLING POLLUTION

30-second foundation

3-SECOND CORE
Future industries will see pollutants as resources that are too valuable to discharge into the natural environment. Engineers can often adapt natural processes to convert pollutants into valuable products.

3-MINUTE IDEA
Government regulations, taxes and incentives to combat pollution enable engineers to develop solutions that provide greater value with stronger community acceptance. As engineers develop cheaper, more efficient techniques to maximize benefits, environmental solutions become profitable, and companies adopt them without needing regulations or incentives.

Engineers have traditionally controlled pollution in two ways: either retaining and storing pollutants until a solution is found; or treating them to an acceptable level before discharging them. Both methods can be expensive, and strong government enforcement is needed for compliance. In developing nations with weak governance, pollutants can often be discharged without significant penalties. Engineers are developing exciting and profitable alternatives, such as cleaner production and industrial ecology. Cleaner production processes designed using 'green chemistry' bypass pollution problems entirely. For example, alumina-refining oxalate residues can be converted into sodium carbonate using bacteria. The sodium carbonate can then be converted into sodium hydroxide to be used in the alumina refining process. Waste containing pollutants from one enterprise can often be converted into valuable products for another. Breweries and food-processing factories generate waste that, instead of being discharged into water, can be converted by bacteria into nutrient-rich fertilizers, generating energy as well as additional income. Engineers adapt natural waste-processing systems for industrial use: man-made swamps with vegetation can be effective waste-processing factories.

RELATED TOPICS
See also
ENVIRONMENTAL ENGINEERING
page 50

RESOURCE SCARCITY
page 142

3-SECOND BIOGRAPHIES

ROBERT UNDERWOOD AYRES
1932–
Formalized industrial ecology concepts.

GATZE LETTINGA
1936–
Developed high-rate anaerobic processes, which inspired contemporary industrial ecologists.

DONALD HUISINGH
1937–
Promotes the ecological modernization movement, arguing that productive use of natural processes can lead to sustainable prosperity.

30-SECOND TEXT
Raj Kurup

Today's pollution is tomorrow's resource, with waste transformed into raw material.

FUTURE TRANSPORT: DRONE SHIPS

30-second foundation

Autonomous or drone ships, operating without a crew, represent the next phase in exploring and navigating the world's oceans. Oceans cover 70 per cent of the Earth's surface, yet 95 per cent remains unexplored. Exploring the ocean is difficult, time-consuming and expensive. At any time, about 160,000 ships are at sea carrying bulk commodities and 5 million containers, but more than 10,000 containers are lost at sea each year. Human error and fatigue are the major causes of maritime accidents. Autonomous ships navigate, monitor their surroundings and detect obstacles using sensors such as cameras, radar, sonar and LiDAR (light detection and ranging). On-board computers interpret the sensor data and control propulsion and steering. Computers use satellite navigation (GPS) and receive weather information along with other ships' locations and identity transmissions. Human captains are still needed, but they will likely be onshore, monitoring multiple autonomous ships, instead of being on board, commanding one ship at sea. Autonomous ships are expected to reduce transportation costs and pollution whilst increasing safety. Other applications include search and rescue, oceanography research, monitoring of dangerous weather and cleaning up pollution.

3-SECOND CORE
Autonomous or drone ships operate without crews using advanced sensors, satellite data and computers to navigate. A land-based captain monitors operations, increasing safety and efficiency.

3-MINUTE IDEA
Although the technology is in place, regulatory changes are needed before autonomous ships can operate in international waters. International law and maritime insurers require that all ships be 'seaworthy'. Definitions of seaworthy demand that ships be staffed by an appropriate crew, shipmaster and pilot. Laws vary from country to country, and also from port to port. Discussions are ongoing to determine how autonomous or drone ships can comply with seaworthiness obligations.

RELATED TOPICS
See also
MECHATRONICS
page 62

ROBOTICS & AUTOMATION
page 70

DRIVERLESS CARS
page 128

3-SECOND BIOGRAPHIES

WILLIAM FROUDE
1810–79
Formulated laws that enabled measurements from small models to be used in the design of full-sized ships.

VICTORIA DRUMMOND
1894–1978
First female member of the Institute of Marine Engineers; oversaw ship building and was honoured for bravery at sea.

MIKAEL MÄKINEN
1957–
President of Rolls-Royce Marine, who has led drone ship technology development.

30-SECOND TEXT
John Krupczak

Unmanned ships can improve sea transport safety and costs and pollution removal.

N
W
HYUNDAI
SIRRAH
UNIFEEDER

APPENDICES

RESOURCES

BOOKS

An Applied Guide to Process and Plant Design
Seán Moran
(Butterworth-Heinemann, 2015)

Built: The Hidden Stories Behind Our Structures
Roma Agrawal
(Bloomsbury Publishing, 2018)

C. Y. O'Connor: His Life and Legacy
A. G. Evans
(University of Western Australia Press, 2001)

Design Paradigms: Case Histories of Error and Judgment in Engineering
Henry Petroski
(Cambridge University Press, 1994)

Disappearing Acts: Gender, Power, and Relational Practice at Work
Joyce K. Fletcher
(MIT Press, 2001)

Educating Engineers
Sheri D. Sheppard, Kelly Macatangay, Anne Colby & William Sullivan
(Jossey-Bass Wiley, 2009)

Electrochemical Science and Technology: Fundamentals and Applications
Keith Oldham, Jan Myland & Alan Bond
(John Wiley & Sons, 2012)

Engineering and the Mind's Eye
Eugene S. Ferguson
(MIT Press, 1992)

Engineering Culture
Gideon Kunda
(Temple University Press, 1992)

Engineering Practice in a Global Context: Understanding the Technical and the Social
Edited by Bill Williams, José Dias Figueiredo & James P. Trevelyan
(CRC/Balkema, 2013)

Environmental and Economic Sustainability
Paul E. Hardisty
(CRC Press, 2010)

Everyday Engineering: An Ethnography of Design and Innovation
Dominique Vinck
(MIT Press, 2003)

Impossible Engineering: Technology and Territoriality on the Canal du Midi
Chandra Mukerji
(Princeton University Press, 2009)

Introduction to Aerospace Materials
Adrian P. Mouritz
(American Institute of Aeronautics, 2012)

Introduction to Biomedical Engineering
John Enderle & Joseph Bronzino
(Academic Press, 2011)

Lees' Loss Prevention in the Process Industries: Hazard Identification, Assessment and Control
Frank Lees
(Butterworth-Heinemann, 2004)

The Making of an Expert Engineer
James P. Trevelyan
(CRC Press/Balkema, 2014)

Materials Selection in Mechanical Design
Michael F. Ashby
(Butterworth Heinemann, 2010)

Notes from Toyota-land: An American Engineer in Japan
Darius Mehri
(Cornell University Press, 2005)

Of Bicycles, Bakelite's, and Bulbs: Toward a Theory of Sociotechnical Change
Wiebe E. Bijker
(MIT Press, 1995)

Railway Transportation Systems: Design, Construction and Operation
Christos N. Pyrgidis
(Taylor & Francis, 2018)

The Reflective Practitioner: How Professionals Think in Action
Donald A. Schön
(Harper Collins, 1983)

Robots for Shearing Sheep: Shear Magic
James P. Trevelyan
(Oxford University Press, 1992)

Soul of a New Machine
Tracey Kidder
(Avon Books, 1981)

Springer Handbook of Nanotechnology
Bharat Bhushan
(Springer-Verlag, 2017)

Springer Handbook of Robotics
Oussama Khatib & Bruno Siciliano
(Springer Verlag, 2016)

Water Engineering
Nazih K. Shammasand & Lawrence K. Wang
(John Wiley & Son, 2016)

WEBSITES

Environmental and Energy Study Institute
www.eesi.org

History of Computers
www.computerhistory.org

International Association for Hydro-Environment Engineering and Research
iahr.org/Web/News_Journals

Software Engineering Introduction
www.edx.org/course/software-engineering-introduction-ubcx-softeng1x

World Federation of Engineering Organizations
www.wfeo.org

NOTES ON CONTRIBUTORS

EDITOR

James Trevelyan is a mechanical engineer and CEO of Close Comfort, developing energy-efficient air conditioners. He led the development of robots for shearing sheep and Internet telerobotics, and developed tools and safety equipment for landmine clearance.

CONTRIBUTORS

Roma Agrawal, MBE, is a structural engineer who has made an impact on the skyline of London with her work on The Shard. She is passionate about promoting engineering careers with under-represented groups.

John Blake is Professor of Engineering and Technology at Austin Peay State University, Tennessee, and advocates a broader understanding of engineering and technology through the development of technological literacy.

Colin Brown is Chief Executive of the Institution of Mechanical Engineers, London, and worked on jet-engine life prediction and led engineering businesses based on advanced materials technologies.

George Catalano is Professor of Biomedical Engineering at Binghampton University, New York, and a former Fulbright scholar and NASA Fellow. He researches aerodynamics, turbulent fluid mechanics and engineering education and ethics.

Doug Cooper is consulting geotechnical engineer with 45 years' experience, and specializes in mine tailings storage design, management and operation.

Kate Disney is Engineering Chair at Mission College, California, has taught a wide range of engineering courses and helps develop a broader understanding of engineering in society.

Roger Hadgraft is a civil engineer and Director for Education Innovation and Research in Engineering at University of Technology Sydney. He is a leading innovator for change towards more practice-based engineering curricula.

Jan Hayes had a long career in chemical engineering safety, and is now Associate Professor in Sociology at RMIT University, Melbourne, specializing in organizational accident prevention.

Marlene Kanga is President of the World Federation of Engineering Organizations, and is a successful innovator, start-up director and chairs the Australian Department of Industry Innovation and Science, R&D Incentives Committee.

Gong Ke is an electronics engineer, President Elect of the World Federation of Engineering Organizations and Professor of Electronic Engineering at Tsinghua, Tianjin and Nankai Universities.

John Krupczak is Professor of Engineering at Hope College, Michigan, and works to improve public understanding about engineering through developing engineering and technological literacy.

Raj Kurup is an environmental consulting engineer, CEO Environmental Engineers International and adjunct professor of University of Missouri and Murdoch University. He has developed economic engineering solutions for waste management and waste water.

Julia Lamborn is Professor of Environmental Engineering at Monash University, Melbourne, and designed and organized the construction of power-station cooling towers for ten years before joining Swinburne University.

Andrew McVeigh is a software engineer with a PhD specializing in the evolution of large systems, and has worked on systems for investment banking, speech synthesis and recognition, video games and in video as the Chief Architect of Hulu.

Seán Moran is a chemical engineer specializing in design, commissioning and troubleshooting of sewage, industrial effluent and water treatment plants.

Paul Newman is BP Professor of Information Engineering at the University of Oxford and Director of the Oxford Robotics Institute, co-founded Oxbotica in 2014, an autonomy vehicle software company.

Hung Nguyen is a researcher in biomedical engineering, artificial intelligence, neurosciences and advanced control. He has developed several biomedical devices and systems for diabetes, disability, cardiovascular diseases and breast cancer.

Jenn Stroud Rossmann is Professor of Mechanical Engineering at Lafayette College, Pennsylvania, is author of the essay series, 'An engineer reads a novel,' combining literary criticism and techno-cultural analysis.

Veena Sahajwalla is Scientia Professor at University of New South Wales, Sydney, and directs the Centre for Sustainable Materials Research and Technology (SMaRT). Her research is advancing sustainability of materials and associated processes in collaboration with industry.

Tomás A. Sancho is civil engineer and general manager of FYSEG, Fulcrum y SERS Engineering Group, Madrid, and has been President of the Ebro Water Confederation and President-founder of three Spanish state water companies.

Jonathan Scott is Foundation Professor of Electrical Engineering at Waikato University, New Zealand, and has expertise in characterization, measurement, modelling and simulation of circuits and systems, especially at radio and microwave frequencies.

Tim Sercombe is materials engineer and Professor and Head of Engineering School at University of Western Australia, and researches additive metal manufacturing techniques for medical implants.

Paul Shearing is Professor of Chemical Engineering at University College London, researches electrochemical technologies and holds the Royal Academy of Engineering Chair in Emerging Battery Technologies.

Donglu Shi researches nanomaterials for energy and medical applications at the University of Cincinnati, Donglu. He is Associate Editor of the Journal of Nanomaterials, and an editorial board member of Nano Research.

Matthew L. Smith is Associate Professor of Engineering at Hope College, Michigan, and researches soft materials that mechanically deform in response to changes in the environment and elastic instabilities in man-made and biological structures.

Jorge Spitalnik is nuclear engineer, past President of World Federation of Engineering Organizations (WFEO) and Executive Director of the Pan American Union of Engineering Societies. He also chaired the WFEO Energy Committee and worked as project manager for ELETRONUCLEAR, Brazil.

Neill Stansbury is a civil engineer and co-founder of GIACC, was Vice Chair of the World Federation of Engineering Organization's Anti-Corruption Committee, past Chair of the International Organization for Standardization (ISO) Anti-Bribery Project Committee and past Chair of British Standards Institution Anti-Bribery Working Group.

INDEX

ACKNOWLEDGEMENTS

The editor would like to thank Sally Male, Roger Hadgraft, John Krupczac and Marlene Kanga for their work as members of the editorial advisory panel for this book.

The publisher would like to thank the following for permission to reproduce copyright material:

Alamy/FALKENSTEINFOTO: 111; FOR ALAN: 47; Iconographic Archive: 65; Newscom: 137; Phillip Harrington: 43; Sueddeutsche Zeitung Photo: 138.
ISO www.iso.org: 23.
Liang Jianying: 118.
Shutterstock/1000 Words: 129; 3DMI: 15; 3drenderings: 99, 135; ADELART: 63; admin_design: 97; AF studio: 141; Africa Studio: 47, 143; akiyoko: 65; Akura Yochi: 67; Alex Farias: 127; Alex Staroseltsev: 45; Alex Tihonovs: 87; alex74: 83; Alexander Lysenko: 27; Alexander Raths: 145; AlexandrBognat: 99; Alexandre Boavida: 105; Alexey Stiop: 51; alexslb: 101; Amovitania: 63; andrea crisante: 17; Andrew S: 77; Andrey Smirnov: 101; Andrey_Kuzmin: 77; Andy Dean Photography: 71; angelh: 79; Angelina Dimitrova: 35; anthonycz: 21; Anton Khrupin: 105; aquatarkus: 143; Arie v.d. Wolde: 77; arslaan: 21; artdig: 43; Artistdesign29: 151; Arun Boonkan: 95; ASDF_MEDIA: 143; athule: 109; attaphong: 27; AUKARAWATCYBER: 149; azure1: 145; BaLL LunLa: 43; Bastian Kienitz: 69; Beautyimage: 51; Bespaliy: 87; Billion Photos: 97; BK Studio: 63; blue-bubble: 51; BlueRingMedia: 105; Botond1977: 101; brovkin: 65; Bubushonok: 77; By Morphart Creation: 111; Camp1994: 47; CapturePB: 97; CGN089: 47; CharacterFamily: 141; cherezoff: 23, 27, 57; Chesky: 135; chinahbzyg: 69; chippix: 25, 141; chor_nat: 77; chris kolaczan: 35; chrisdorney: 35; CK Foto: 149; corlaffra: 83; Creative Stall: 71; Creative Thoughts: 137; CSLD: 57; DAMRONG RATTANAPONG: 143; Darkydoors: 45; Dawid Lech: 41; Dennis Jacobsen: 107; dimitris_k: 85; Dmytro Bochkov: 99; Dn Br: 111; DOCTOR BLACK: 97; dominika zarzycka: 105; Dominique Mills: 121; donatas1205: 143; Donna Beeler: 25; Dotted Yeti: 127; Dr. Norbert Lange: 149; DyziO: 49; Dzmitry_Kuzniatsou: 121; Elzbieta Sekowska: 99, 143; eNjoy iStyle: 87; enjoy your life: 35, 43; Ensuper: 99; ermess: 127; Evelia Smith: 81; Everett Collection: 49, 61, 109; Everett Historical: 147; Ezume Images: 27; Fabian Fuenmayor Bernal: 95; Fine Art: 129; Flipser: 99; Fotokon: 83; FrameStockFootages: 85; franco lucato: 141; Frannyanne: 87; FUN FUN PHOTO: 129; galimovma79: 63; Galina 2703: 69; geen graphy: 101; Giamportone: 141; gillmar: 95; GLF Media: 129; gritsalak karalak: 127; gyn9037: 151; HacKLeR: 25; Hal_P: 63; Happy_stocker's: 145; Hein Nouwens: 45, 135; Hekla: 135; HelloRF Zcool: 61; Hemerocallis: 87; HM Design: 123; HodagMedia: 19; hramovnick: 85; iadams: 123; iamvlsv: 151; Igor Kyrlytsya: 79; Illerlok_Xolms: 141; Inked Pixels: 57, 105; Ioan Panaite: 107; IR Stone: 107; Ivan Cholakov: 97; Iynea: 15; Jakub Krechowicz: 137; Jantana: 45; jassada watt_: 63; javarman: 51; JFs Pic S. Thielemann: 57; Joe Dejvice: 67; jordeangjelovik: 109; josh.tagi: 79; JPC-PROD: 87; Julia_Ko: 97; kakteen: 97; Kao-len: 47; Karrrtinki: 137; Kateryna Kon: 105; Katherine Welles: 145; Ken StockPhoto: 135; kezza: 147; Khadi Ganiev: 41; Kiattipum: 99; kilic inan: 81; Kindlena: 63; Kletr: 85; Kniazeff: 23; komkrit Preechachanwate: 25; konstantinks: 81; Konstantynov_AA: 121; koya979: 137; KPG Ivary: 109; Krasowit: 137; KREML: 123; Ksenia_designer: 41; Laboko: 145; LDEP: 123; LesPalenik: 143; Leszek Glasner: 47; Liu zishan: 23; lmstockwork: 63; Luciano Cosmo: 85; M.Fuksa: 79; Makhnach_S: 57; MarcelClemens: 15; Masson: 85; Mauvries: 41; Maxx-Studio: 43; Mega Pixel: 27; MIA Studio: 137; Michal Ninger: 41; MidoSemsem: 137; Mike Russell: 117; miri019: 149; MISS KANITHAR AIUMLA-OR: 95; Monkey Business Images: 21; Morphart Creation: 41, 45, 135; Mrs_ya: 61, 125; mrwebhoney: 137; MSSA: 99; muratart: 117; muuraa: 61; MyImages – Micha: 19; Natursports: 45; Natykach Nataliia: 151; NavinTar: 77; Neale Cousland: 47; Nel4: 17; Nerthuz: 81, 109; ngaga: 141; nimon: 49; nitinut380: 17; nogreenabovetwothousand: 117; NoPainNoGain: 87; Noppharat4969: 49; NotarYES: 97; nowment: 129; Nuno Andre: 123; Nuttawut Uttamaharad: 125; Olaf Naami: 63; Oleksii Lishchyshyn: 151; Only background: 107, 123; OPgrapher: 87; OvuOng: 147; Panda Vector: 121; paniti Alapon: 109; Pavlo Lys: 87; pedrosala: 117; petovarga: 141; philia: 109; PhilMacDPhoto: 69; phipatbig: 71; Phongsak Meedaenphai: 107; Phonlamai Photo: 71, 105; Photobank gallery: 107; Photomontage: 71; PhotoSky: 141; Podsolnukh: 67; POM POM: 97; Potapov Alexander: 49, 51; Praphan Jampala: 23; Preto Perola: 27; Production Perig: 49; psynovec: 45; ptashka: 51; pukach: 105; Quatrox Production: 63; r.classen: 39; railway fx: 39; RasaSopittakamol: 145; Rashad Ashur: 19; Rasica: 121; Rawiz: 43; Rawpixel.com: 107; Razvan Dan Paun: 123; Repina Valeriya: 141; Rich Carey: 87; Richman Photo: 111; Robert Kneschke: 43; robertwcoy: 95; Robsonphoto: 79; Rolaks: 63; Roman Sigaev: 81; rommma: 57; ronstik: 61; Rungruedee: 111; sarawut muensang: 147; Sashkin: 19; SAYAM TRIRATTANAPAIBOON: 97; schankz: 51; Sebastian Kaulitzki: 111; sebra: 25; Sergey Nivens: 137; Sergieiev: 19; Sergio Stakhnyk: 145; Serhii Bobyk: 61; Serjio74: 117; shaineast: 65; Simon Annable: 47; SimonWest: 35; Skycolors: 65; smile plus: 41; Somchai Som: 79; Sonpichit Salangsing: 83; Soon Wee hong: 57; spainter_vfx: 101; StanislauV: 15; stefan11: 79; stefanphotozemun: 125; Stephen Finn: 49; STILLFX: 99; Stocksnapper: 61; Strejman: 95; struvictory: 117; studiolab: 145; studiovin: 141; Sulee_R: 149; Suwin: 79; SvitlanaNiko: 99, 151; Svjatoslav Andreichyn: 39; Sylfida: 63; Sylverarts Vectors: 17; T. Lesia: 125; tackune: 117; tammykayphoto: 123; Tampo: 143; tankist276: 121; technobulka: 125; think4photop: 83; tickcharoen04: 69; Tony Baggett: 35; Tumarkin: 107; Uladzislau Baryschchyk: 151; Ume illus4289: 51; Uncimo: 101; Vacancylizm: 83; Vadim Georgiev: 25; VanderWolf Images: 65; Varunyuuu: 39; vberla: 137; Vector FX: 21; VectorA: 21; VectorShow: 67; Viktor Shumatov: 129; Vintage Tone: 63; Vlad Kochelaevskiy: 15; Vladi333: 127; Vladimir Gjorgiev: 137; vonzur: 51; VPales: 81; Vyntage Visuals: 21, 87; wavebreakmedia: 49; WEB-DESIGN: 141; WERAYUTH PIRIYAPORNPRAPA: 41; Xenia Design: 51; yauhenka: 21; Yellow Cat: 61; yevgeniy11: 95; YoPixArt: 135; Yuri Schmidt: 77; zcw: 61; Zelenskaya: 41; Zonda: 61
Wikimedia Commons/Benjamin Baker: 39; Carolus-Duran: 35; Garitzko: 151; Georges Darrieus - US Patent 1835018: 69; Great Western Railway/ Geof Sheppard Collection: 35; Illustration from 1911 Encyclopaedia Britannica, article Bridges: 49; James S. Davis: 102; State Library of Western Australia: 28; The Nobel Foundation, BASF – Pressephotos: 88; Theo's Little Bot: 36; Thomas Annan: 58.

All reasonable efforts have been made to trace copyright holders and to obtain their permission for the use of copyright material. The publisher apologizes for any errors or omissions in the list above and will gratefully incorporate any corrections in future reprints if notified.